P9-DTY-550

TPM DEVELOPMENT PROGRAM

Implementing Total Productive Maintenance

TPM DEVELOPMENT PROGRAM

Implementing Total Productive Maintenance

Edited by Seiichi Nakajima

Introduction by Benjamin S. Blanchard
Virginia Polytechnic Institute

Originally published by the Japan Institute for Plant Maintenance

Productivity Press
Cambridge, Massachusetts   Norwalk, Connecticut

Originally published as *TPM tenkai* by the Japan Institute for Plant Maintenance, Tokyo, Japan. Additional copies of this book are available from the publisher. Address all inquiries to:

Productivity Press, Inc.
P.O. Box 3007
Cambridge, MA 02140
(617) 497-5146

Library of Congress Catalog Card Number: 88-43566
ISBN: 0-915299-37-2

Cover design by Gail Graves
Set in Palatino and Helvetica by Rudra Press, Cambridge, MA
Printed and bound by The Maple-Vail Book Manufacturing Group
Printed in the United States of America

Library of Congress Cataloging-in-Publication Data

TPM tenkai. English.
TPM development program: implementing total productive maintenance/edited by Seiichi Nakajima; introduction by Benjamin S. Blanchard; contributors, Seiichi Nakajima . . . [et al.].
Translation of: TPM tenkai.
1. Plant maintenance—Management. I. Nakajima, Seiichi, 1928- .
II. Title. III. Title: TPM development program.
TS192.T6813 1989 88-43566 658.2'02—dc20 CIP
ISBN 0-915299-37-2

89 90 91 92 10 9 8 7 6 5 4 3 2 1

Table of Contents

Publisher's Foreword

In the ideal factory, equipment should be operating at 100 percent capacity 100 percent of the time. TPM is a powerful concept leading us close to the ideal with no downtime, no defects, and no safety problem. This book, *TPM Development Program,* offers a detailed explanation of the integrated spheres of improvement activity that we now recognize as TPM: equipment improvement, autonomous maintenance, skills training for operators and maintenance personnel, improved maintenance management, and maintenance prevention activities.

In these contexts, if total productive maintenance is a companywide approach to quality through equipment, what does "total" mean? Certainly, it means involving everyone in the company in the programs. It also means transferring knowledge — a certain kind of understanding — to everyone. TPM expands the knowledge base of operators and maintenance personnel and brings them together as a cooperative team to optimize PM activities. TPM should also be a major issue for senior management, however, because eliminating breakdowns and reducing defects-in-process also promotes the successful achievement of a streamlined continuous flow process. *In other words, optimally functioning equipment is a vital piece in the just-in-time puzzle.*

Taiichi Ohno, originator of the Toyota production system, often speaks of the ideal production system as a sensitive flow, like the nervous system of the human body. To act and react efficiently, each part of the production system (like each part of the body) must be sensitive to external stimuli and able to respond to new demands immediately. This means shared understanding, effective communication, and coordinated action — themes that are repeatedly underscored in *TPM Development*. Improvements emerging from isolated pockets of enthusiasm on the shop floor or among a few engineers may ultimately result in overall disappointment in a TPM program. Therefore, at every stage of TPM implementation explored in this book, new ways of thinking, communicating, and working are emphasized.

Teamwork and small group activity, for example, are important at all levels and at every stage, because they promote the flow of shared knowledge and information — the TPM "equipment consciousness" and the transfer of new data gained through a "shop-floor approach" to problem-solving. For example, operators understand and begin to interact with their equipment; their involvement in daily PM in turn allows maintenance personnel to expand their own improvement horizons; they have more time to improve equipment maintainability, to learn about and use more sophisticated techniques.

Communication is vital and is continuously stressed. A good example is maintenance-prevention cross-functional team activities, which represent a potentially powerful collaboration between maintenance, engineering, and design staff. Data derived from day-to-day observation of equipment is translated into information that can be used to design new, virtually maintenance-free equipment or to make wiser equipment purchases. Enhanced communication between these groups also reduces time and eliminates costly delays to correct errors and unanticipated problems between the design and commercial operation stages.

Another important key in TPM is the combination of a zero-defects philosophy with a shop-floor team approach to equipment problem-solving. Dealing with persistent, chronic losses forces team members to put aside narrow engineering solutions or approaches that worked in the past for innovative, experimental, hands-on work based on observation of the actual equipment.

Every participant benefits from these activities and enhanced communication systems document and help to share the valuable insights with other teams working on similar problems.

TPM Development Program was written by a team of consultants at the Japan Institute of Plant Maintenance and is rich in insights, case examples, and gritty truths about the work of improving equipment and maintenance systems. It follows *Introduction to TPM*, (Cambridge: Productivity Press, 1988), in what we hope will be a valuable series of books and products on this important subject. Our special thanks go out to Tajiri Masaji of JIPM who, on behalf of Mr. Nakajima, devoted hundreds of hours and a week-long visit at his own expense to help us provide a correct and complete translation. Finally, thanks to Esmé McTighe, who produced the book with the able assistance of designer Donna Puleo and our friends at Rudra Press, Caroline Kutil, Michele Seery, Gayle Joyce, and Susan Cobb.

Norman Bodek
President
Connie Dyer
Senior Editor

Preface to English Edition

What is the secret of Japanese quality and productivity? For some time, the standard answer to this question has been just-in-time production (JIT), the uniquely Japanese production system developed at Toyota Motor Company, and TQC, the Japanese approach to total quality control.

Excellent Japanese companies have another secret, however, that has pushed productivity and quality to their limits, making possible production lines with zero breakdowns and zero defects. That secret is TPM, or total productive maintenance.

Only a few companies in the world are aware of TPM, perhaps because so little has been published in English on the subject. Now, however, Productivity Press has produced an English-language version of *TPM Development Program*, an in-depth follow-up volume to *Introduction to TPM*, published in English in the summer of 1988.

Introduction to TPM is intended to offer an executive's summary of TPM. It outlines the principles and programs of TPM for anyone willing to take an evening to read it. *TPM Development Program*, on the other hand, provides a closer, more detailed view of the various TPM programs, illustrated by case materials from many cooperating companies. These two books have served together as the "TPM bible" for Japanese companies developing their own TPM programs and have been reprinted in Japan many times since their initial publication five or six years ago. We expect that publication of these new English-language editions will contribute to wider understanding that TPM, like JIT and TQC, is a fundamental factor in world-class quality and productivity.

Although the basic TPM concepts and programs have changed very little since their introduction, TPM has been implemented in many different industries. For example, while most examples in the book reflect the experiences of PM prizewinners in the machining and assembling industries, today TPM is being applied increasingly in process industries.

TPM Development Program, read together with its companion, *Introduction to TPM,* should serve as an excellent basis for understanding total productive maintenance. We encourage companies to pursue the development of TPM in their own factories and to experience the improvements in quality and productivity that it makes possible.

Finally, on behalf of the authors, I would like to express our gratitude to all those at Productivity Press who tackled the English translation of this book so enthusiastically.

Seiichi Nakajima, Editor
Vice Chairman, Japan Institute of Plant Maintenance

Introduction

In today's environment, systems are becoming increasingly complex and their performance and effectiveness are often inadequate to meet consumer needs. The costs associated with their acquisition and use are also growing rapidly. At the same time, competition is increasing; there is a greater degree of international cooperation and exchange, and the requirements for producing a well-integrated, cost-effective system are even greater than in the past. System consumers are demanding higher quality and more cost-effective results whether they are buying a manufacturing capability, an electronics system, or an automobile.

From an economic standpoint, *total cost visibility* is often lacking. While we have been quite successful in dealing with the short-term aspects of cost, we have been less responsive to long-term effects. For example, design and development costs and the costs of acquiring and installing a system are often relatively well known, while much of the cost associated with its operation and maintenance is hidden. As we have learned, the percentage of a system's life cycle cost attributable to operational and maintenance activities can be quite large — up to 75 percent for some systems. Obviously, these costs will have a negative impact on profits, and in a highly competitive environment, they can affect product sales. This is particularly true in manufacturing, where consumer product prices must be adjusted upward to cover high maintenance costs.

Looking for the cause, we find that the major portion of a system's projected life cycle cost stems from decisions made during the early phases of advanced planning and conceptual design. Decisions relating to system operation, maintenance and support policies, manual versus automated applications, equipment selection, packaging-design schemes, level of repair concepts, and so on, have a great impact on life cycle cost. In other words, the greatest opportunity for reducing life cycle cost, a large percentage of which is attributed to system operation and maintenance, exists during the early phases of system development.

Additionally, trends in today's environment, combined with the challenge of intense international competition, increasingly require us to deal with systems in *total*, projected in terms of their *life cycle* and developed on a well-integrated, *top-down* basis.

The system includes not only the equipment directly involved in the manufacture of a consumer product but the maintenance support capability as well. We must consider the need for maintenance personnel, test equipment and tools, spare parts, special facilities, data, and software, along with production facilities, the prime equipment, and operating personnel. These elements must be integrated and balanced with the other components of the system, and this must be accomplished early in the system development process.

Once specific qualitative and quantitative requirements for the overall system are defined, they must be allocated in terms of design criteria; trade-off analyses and design optimization are then accomplished, followed by system development for consumer use. This process, which must include maintenance considerations, makes it possible to address the system (and its components) as an entity, and to do so early in the life cycle, when the costs of incorporating possible changes are minimized.

As for "high-cost contributors" to life cycle cost, experience indicates that system maintenance has been a major cause! Many of the costs associated with maintenance have been due to equipment breakdowns, the continued operation of degraded equipment, the inadequacies of maintenance personnel, and the unavailability of spare parts, test equipment, data, and so on. These factors lead to unnecessary downtime, production losses, and the waste of valuable resources.

Often, these costs are the result of poor equipment design decisions such as selecting unreliable or unmaintainable equipment. In other cases, such costs result from a poorly designed support capability — for example, lack of properly trained maintenance personnel or, simply, the lack of appropriate spare parts when needed.

The total productive maintenance (TPM) development program elaborated in this book and in the earlier *Introduction to TPM* meets many of these concerns directly. First, the high contribution of operational and maintenance activities to life cycle cost is reduced through participative programs designed to increase equipment effectiveness. These programs call for

- group activities to eliminate the six major equipment-related losses
- restoration of all equipment to optimal operating conditions and elimination of accelerated deterioration
- the involvement of operators in daily autonomous maintenance activities to maintain basic equipment conditions (daily inspection, cleaning, lubrication, and bolting)
- maintainability improvements for existing equipment
- increased efficiency and cost-effectiveness of maintenance work through better scheduling and management

Second, the impact of early planning and design-stage decisions on life cycle cost is addressed in the TPM maintenance-prevention programs. By promoting an integrated total-system approach and close cooperation between maintenance, engineering, and design engineers, valuable maintenance data in a usable form becomes available in the earliest stages, thus promoting maintenance-free design in new equipment. Moreover, the same cooperative approach is taken to systematic debugging and correction of design weaknesses before the commissioning and startup stages, when problems are more costly to correct and are often treated superficially.

In essence, system maintenance must be addressed on a basis comparable with system performance if we are to truly meet consumer demands in today's environment. Reliability, maintainability, and supportability parameters must be incorporated into the earliest phases of design; the appropriate trade-off analyses must be accomplished to ensure the proper balance between corrective maintenance and preventive maintenance; an effective maintenance capability must be established throughout the consumer-use phase to ensure high-quality support; and a data collection, analysis, and feedback mechanism must be incorporated for the purposes of initiating corrective action or modifications for system improvement. Maintenance must be addressed in the context of the *total* system, on an *integrated* basis, and in terms of the *life cycle*.

Fulfillment of these objectives requires the implementation of an effective, integrated maintenance management program. I believe that the TPM concept presented in this text is excellent, and that a program designed with TPM as the major thrust should permit any large-scale system to achieve cost-effectiveness goals.

The authors do an excellent job of introducing TPM, describing the various categories of maintenance and some of the relationships between maintenance and system quality, and identifying the major steps required in implementing a successful maintenance program. Of particular significance is the approach used to involve managers and employees as partners in the accomplishment of maintenance program objectives. Finally, I believe that the concepts discussed are applicable to any type of system, even though the emphasis here is on the production process; that is, the factory environment.

I wish to express my appreciation to Mr. Seiichi Nakajima, the authors of this text, and the Japan Institute for Plant Maintenance (JIPM) for allowing me the opportunity to review and comment on this excellent exposition of a TPM development program.

Benjamin S. Blanchard
Assistant Dean and Professor
College of Engineering
Virginia Polytechnic Institute and State University

TPM DEVELOPMENT PROGRAM

Implementing Total Productive Maintenance

1

An Introduction to TPM

After World War II, Japanese industries determined that to compete successfully in the world market they had to improve the quality of their products. To do so, they imported management and manufacturing techniques from the United States and adapted them to their circumstances. Subsequently, their products became known throughout the world for their superior quality, focusing world attention on Japanese-style management techniques.

FROM PREVENTIVE MAINTENANCE (PM) TO TOTAL PRODUCTIVE MAINTENANCE (TPM)

To improve equipment maintenance, Japan imported the concept of preventive maintenance (PM) from the United States more than thirty years ago. Later imports included productive maintenance (also known as PM), maintenance prevention (MP), reliability engineering, and so on. What is now referred to as TPM is, in fact, American-style productive maintenance, modified and enhanced to fit the Japanese industrial environment.

In most American companies, maintenance crews perform all factory maintenance, enforcing an "I operate — you fix" division of labor. By contrast, many Japanese corporations have modified American PM so that all employees can participate. Total productive maintenance (TPM), often defined as productive maintenance implemented by all employees, is based on the principle that equipment improvement must involve everyone in the organization, from line operators to top management.

The key innovation in TPM is that operators perform basic maintenance on their own equipment. They maintain their machines in good running order and develop the ability to detect potential problems before they generate breakdowns.

TPM was introduced in Japan more than ten years ago and has since found wide acceptance. For example, TPM provides essential support for the Toyota production system. TPM has also been implemented by many Toyota-affiliated companies. According to its creator, Taiichi Ohno, the Toyota production system is based on the absolute elimination of waste. In Toyota's just-in-time production, only "the necessary items are produced, when needed, and in the amounts needed." In other words, the Toyota production system strives to attain zero defects and zero inventory levels.

Figure 1-1 illustrates the relationship between TPM and the basic features of the Toyota production system. As this matrix shows, the purpose of TPM is to eliminate the six big losses, which corresponds to Toyota's absolute elimination of waste.

In striving for zero breakdowns, TPM promotes defect-free production, just-in-time production, and automation. Without TPM, the Toyota production system could probably not function. This is confirmed by the fact that Toyota affiliates are rapidly implementing TPM.

Nippondenso Co., a well-known Toyota supplier of electrical parts, began implementing productive maintenance in 1961. In 1969, to keep up with the rapid progress in automated production, the firm successfully implemented "productive maintenance with total employee participation" (TPM). Two years later, it was the first company to be awarded the Distinguished Plant Prize (PM Prize) for its achievements with TPM. Since then, the PM Prize has been awarded annually on the basis of TPM implementation.*

---

* The Japan Institute for Plant Maintenance (JIPM) has been awarding the PM Prize since 1964 (*see* Appendix A, p.377). In making its annual awards, the PM Prize Committee focuses on actual TPM effectiveness, based on complete elimination of equipment losses, increased productivity, better quality, reduced costs, minimal inventory, elimination of accidents and pollution, and a pleasant working environment.

TPS \ TPM	Breakdowns	Setup and adjustment	Idling and minor stoppage	Reduced speed	Quality defects	Reduced yield (from startup)
Implementing flow process	●					
Eliminating defects					●	●
Stockless production	●	●				
Reduced lot size		●				
Quick setup		●				
Standard cycle times	●	●	●	●	●	
Standard production sequence	●	●	●	●	●	
Standard idle time	●	●	●	●	●	
Visual control *andon* line-stop alarm	●	●	●			
Improved machine operability	●	●				
Improved maintainability	●					

Figure 1-1. Toyota Production System and TPM

History of TPM

Preventive maintenance was introduced in the 1950s and productive maintenance became well-established during the 1960s (Table 1-1). The development of TPM began in the 1970s. The time prior to 1950 can be referred to as the "breakdown maintenance" period.

	1950s	1960s	1970s
Era	**Preventive Maintenance** Establishing maintenance functions	**Productive Maintenance** Recognizing importance of reliability, maintenance, and economic efficiency in plant design	**Total Productive Maintenance** Achieving PM efficiency through a comprehensive system based on respect for individuals and total employee participation
Approach	• PM (Preventive Maintenance) 1951- • PM (Productive Maintenance) 1954- • MI (Maintainability Improvement) 1957-	• Maintenance prevention 1960- • Reliability engineering 1962- • Maintainability engineering 1962- • Engineering economy	• Behavioral sciences • MIC, PAC, and F plans[1] • Systems engineering • Ecology • Terotechnology • Logistics
Major Event	**1951** Toa Nenryo Kōgyō is the first Japanese company to use American-style PM **1953** 20 companies form a PM research group (later the Japan Institute of Plant Maintenance (JIPM) **1958** George Smith (U.S.) comes to Japan to promote PM	**1960** First maintenance convention (Tokyo) **1962** Japan Management Association sends mission to U.S. to study equipment maintenance **1963** Japan attends international convention on equipment maintenance (London) **1964** First PM Prize awarded in Japan **1965** Japan attends international convention on equipment maintenance (New York) **1969** Japan Institute of Plant Engineers (JIPE) established	**1970** International convention on equipment maintenance held in Tokyo (co-sponsored by JIPE and JMA) **1970** Japan attends international convention on equipment maintenance sponsored by UNIDO[2] (West Germany) **1971** Japan attends international convention on equipment maintenance (Los Angeles) **1973** UNIDO sponsors mainte-nance repair symposium in Japan **1973** Japan attends international terotechnology convention (Bristol, England) **1974** Japan attends EFNMS[3] maintenance congress **1976** Japan attends EFNMS maintenance congress **1978** Japan attends EFNMS maintenance congress **1980** Japan attends EFNMS maintenance congress

[1] Management for Innovation and Creation (MIC); Performance Analysis and Control (PAC); Foreman Plan (F Plan)
[2] United Nations Industrial Development Organization (UNIDO)
[3] European Federation of National Maintenance Societies (EFNMS)

Table 1-1. History of PM in Japan

Category	Examples of TPM Effectiveness
P (Productivity)	• Labor productivity increased: 140% (Company M) 150% (Company F) • Value added per person increased: 147% (Company A) 117% (Company AS) • Rate of operation increased: 17% (68% → 85%) (Company T) • Breakdowns reduced: 98% (1,000 → 20 cases/mo.) (Company TK)
Q (Quality)	• Defects in process reduced: 90% (1.0% → 0.1%) (Company MS) • Defects reduced: 70% (0.23% → 0.08%) (Company T) • Claims from clients reduced: 50% (Company MS) 50% (Company F) 25% (Company NZ)
C (Cost)	• Reduction in manpower: 30% (Company TS) 30% (Company C) • Reduction in maintenance costs: 15% (Company TK) 30% (Company F) 30% (Company NZ) • Energy conserved: 30% (Company C)
D (Delivery)	• Stock reduced (by days): 50% (11 days → 5 days) (Company T) • Inventory turnover increased: 200% (3 → 6 times/mo.) (Company C)
S (Safety/ Environment)	• Zero accidents (Company M) • Zero pollution (every company)
M (Morale)	• Increase in improvement ideas submitted: 230% increase (36.8/person → 83.6/person) (Company N) • Small group meetings increased: 200% (2 → 4 meetings/mo.) (Company C)

ble 1-3. Examples of TPM Effectiveness (Recipients of the PM Prize)

Japanese companies have implemented TPM in stages roughly corresponding to the stages of PM development in Japan between 1950 and 1980 (Table 1-2). The information in Table 1-2 is based on data collected in 1976 and 1979 from 124 plants belonging to the JIPM. In three years, the number of plants actively practicing TPM more than doubled. Now, over 20 percent of these factories practice TPM.

		1976	1979
Stage 1	Breakdown maintenance	12.7%	6.7%
Stage 2	Preventive maintenance	37.3%	28.8%
Stage 3	Productive maintenance	39.4%	41.7%
Stage 4	TPM	10.6%	22.8%

Table 1-2. Four Stages of PM Development and the Current Situation in Japan

TPM and the Future of Maintenance

Until the 1970s, PM in Japan consisted mainly of preventive, or time-based maintenance, featuring periodic servicing and overhaul. During the 1980s preventive maintenance is rapidly being replaced by predictive, or condition-based maintenance. The success of TPM depends on our ability to be continuously aware of the condition of equipment in order to predict (and prevent) failures. Predictive maintenance plays a significant role in TPM, because it uses modern monitoring techniques to diagnose the condition of equipment during operation by identifying signs of deterioration or imminent failure.

HOW DOES TPM WORK?

TPM is productive maintenance carried out by all employees through small group activities. Like TQC, which is companywide total quality control, TPM is equipment maintenance performed on a companywide basis. The term TPM was defined in 1971 by the Japan Institute of Plant Engineers (forerunner of

Japan Institute for Plant Maintenance) to include the following five goals:

1. Maximize equipment effectiveness (improve overall efficiency).
2. Develop a system of productive maintenance for the life of the equipment.
3. Involve all departments that plan, design, use, or maintain equipment in implementing TPM (engineering and design, production, and maintenance).
4. Actively involve all employees — from top management to shop-floor workers.
5. Promote TPM through *motivation management*: autonomous small group activities.

The word "total" in "total productive maintenance" has three meanings related to three important features of TPM:

- *Total effectiveness*: pursuit of economic efficiency or profitability
- *Total PM*: maintenance prevention and activity to improve maintainability as well as preventive maintenance
- *Total participation*: autonomous maintenance by operators and small group activities in every department and at every level

The first concept, *total effectiveness* (or "profitable PM"), is emphasized in predictive and productive maintenance (Figure 1-2).

The second concept, *total PM*, was also introduced during the productive maintenance era. It means establishing a maintenance plan for the entire life of the equipment and includes maintenance prevention (MP: maintenance-free design), which is pursued during the equipment design stages. Once equipment is installed, a total maintenance system requires preventive maintenance (PM: preventive medicine for equipment) and maintainability improvement (MI: repairing or modifying equipment to prevent breakdowns and facilitate ease of maintenance).

The last concept, *total participation*, which includes autonomous maintenance by operators and small group activities, is unique to TPM.

	TPM features	Productive Maintenanc features
Economic efficiency (profitable PM)	○	○
Total system (MP-PM-MI)*	○	○
Autonomous maintenance by operators (small group activities)	○	

TPM = Productive Maintenance + small-group activities

*MP = maintenance prevention
PM = preventive maintenance
MI = maintainability improvement

Figure 1-2. Relationship Between TPM, Productive Maintenance and Preventive Maintenance

Examples of TPM Effectiveness

TPM has a double goal — zero breakdowns fects. When breakdowns and defects are eliminate operation rates improve, costs are reduced, inv minimized, and as a consequence, labor producti As Table 1-3 illustrates, one firm reduced the nur downs to 1/50 of the original number; some compan 26 percent increases in equipment operation rates show a 90 percent reduction in process defects; la ity generally increased by 40 to 50 percent.

Of course, such results cannot be achieved o cally, three years are required from the introduct achieve prizewinning results. Furthermore, in the TPM, the company must bear the additional exper equipment to its proper condition and educat about the equipment. The actual cost depends on ity of the equipment and the technical expertise a of maintenance staff. As productivity increases, l costs are quickly recouped. This is why TPM is o as "profitable PM."

Maximizing Equipment Effectiveness

The goal of all factory improvement activity is to increase productivity by minimizing input and maximizing output. *Output* refers not only to increased productivity but also to better quality, lower costs, timely delivery, improved industrial safety and hygiene, higher morale, and a more favorable working environment.

The relationship between input and output can be visualized as a matrix (Figure 1-3). Human workers, machinery, and material are combined as *input*, while *output* consists of PQCDSM — production (P), quality (Q), cost (C), delivery (D), safety, hygiene, and environment (pollution-control) (S), and morale or human relations (M). The right column lists the method by which each output factor is regulated. The input factors are determined by manpower allocation, plant engineering and maintenance, and inventory control.

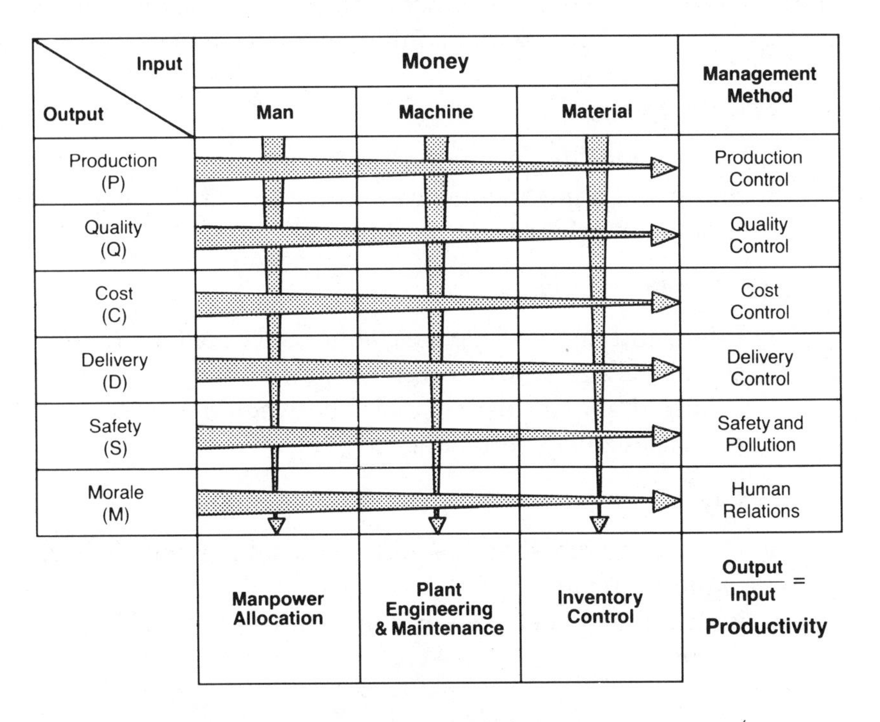

Figure 1-3. Relationship Between Input and Output in Production Activities

This matrix makes it obvious that plant engineering and maintenance are directly related to all output factors (PQCDSM). With increased automation and reduction of labor, production shifts from the hands of the workers to the machinery. At this point, equipment and machinery are the crucial factors in increasing output. Productivity, quality, cost, and delivery, as well as safety, hygiene, environment, and morale are all influenced significantly by equipment conditions.

The goal of TPM is to enhance equipment effectiveness and maximize equipment output (PQCDSM). It strives to attain and maintain optimal equipment conditions in order to prevent unexpected breakdowns, speed losses, and quality defects in process. Overall efficiency, including economic efficiency, is achieved by minimizing the cost of upkeep and maintaining optimal equipment conditions throughout the life of equipment, in other words, by minimizing life cycle cost (LCC).

Equipment effectiveness is maximized and life cycle cost minimized through companywide efforts to eliminate the following "six big losses" that reduce equipment effectiveness:

Downtime

1. Breakdowns due to equipment failure
2. Setup and adjustment (*e.g.*, exchange of die in injection molding machines, etc.)

Speed losses

3. Idling and minor stoppages (abnormal operation of sensors, blockage of work on chutes, etc.)
4. Reduced speed (discrepancies between designed and actual speed of equipment)

Defects

5. Defects in process and rework (scrap and quality defects requiring repair)
6. Reduced yield between machine startup and stable production

The Relationship Between TPM, Terotechnology, and Logistics

To the extent that TPM focuses on reducing life cycle costs, it has features in common with the concepts of terotechnology, developed in the United Kingdom, and logistics.

Terotechnology is a term coined in the United Kingdom in 1970. It combines management, financial, engineering, and other practices with physical assets to achieve economic life cycle costs (LCC). It is concerned with the specification and design for reliability and maintainability of plant machinery and structures; with their installation, commissioning, maintenance, modification, and replacement; and with feedback of information on design, performance, and costs.

Focusing only on equipment, TPM has the same goal as terotechnology — economic life cycle cost. Actually, the United States Department of Defense first articulated the concept of economic life cycle cost in 1965, when it issued MIL-STD-785 — *Reliability Programs for Systems and Equipment* — which mandated the integration of reliability engineering activities with traditional engineering activities in design, development, and production to eliminate potential reliability problems at the earliest and cheapest stages in the development cycle. Since 1976, the department has based its procurement contracts for weapons and other large-scale systems on LCC.

Logistics is a term borrowed from the military, meaning aid to the front line through procurement, storage, transportation, and maintenance of manufactured goods and systems. This discipline has modernized traditional notions of the life cycle of goods and equipment through the concepts of LCC, reliability engineering, and maintenance engineering.

Although it is true that TPM, terotechnology, and logistics have economic LCC as a common goal, they differ in terms of the precise target and the assignment of responsibility (Figure 1-4). Logistics targets an extremely wide field, including manufactured goods, systems, programs, information, and equipment.

Focusing only on equipment (available assets), terotechnology involves the equipment supplier, engineering firms, and the equipment user. TPM, on the other hand, is practiced only by the equipment user.

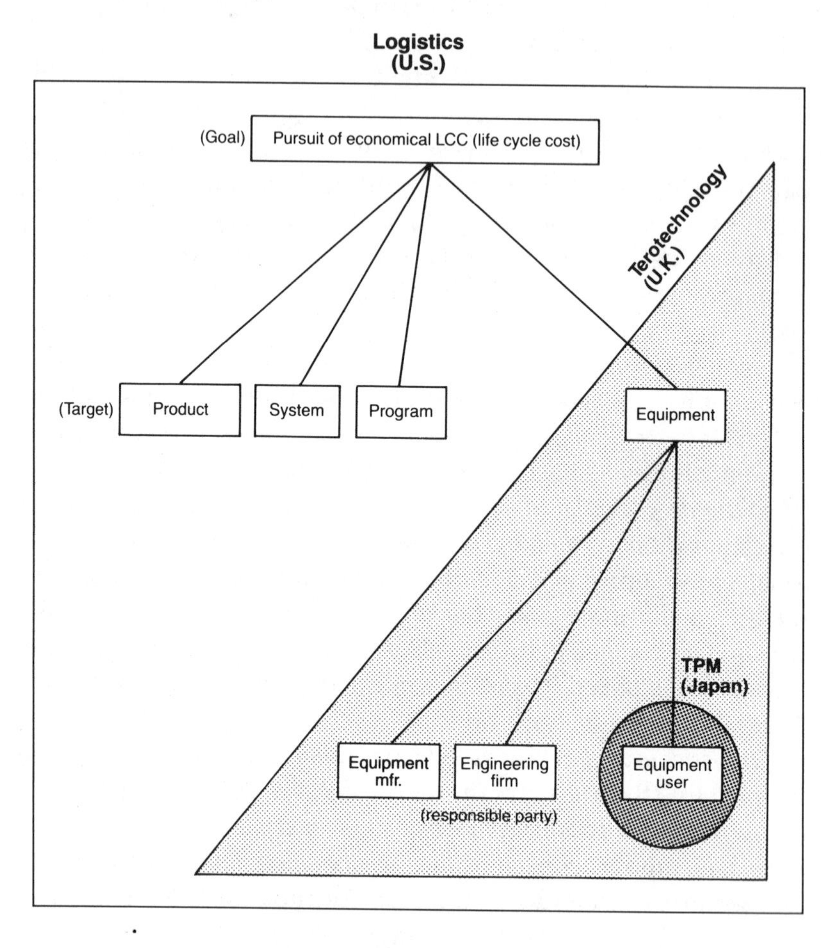

Figure 1-4. TPM, Terotechnology, and Logistics

INTRODUCING TPM INTO THE FACTORY

In Japan, three major factors in workplace improvement are *yaruki* (motivation), *yaruude* (competence), and *yaruba* (work environment). TPM addresses all three: it leads to fundamental corporate improvement by improving worker and equipment utilization. To eliminate the six big losses we must first change people's attitudes or motivation (*yaruki*) and increase their skills (*yaruude*). We must also create a work environment (*yaruba*) that supports TPM implementation. Unless top management takes the lead by tackling this issue, however, the necessary transformation in attitudes, equipment, and the corporate environment for change will not progress smoothly.

TPM as Basic Company Policy

TPM combines "top-down" goal-setting by top management with "bottom-up" small-group improvement and maintenance activities on the front line. Top management must incorporate TPM into the basic company policy and establish concrete goals, such as increasing the rate of equipment operation to more than 80 percent or reducing breakdowns by 50 percent over the course of several years. TPM can succeed only with the commitment of top management — if managers are determined to implement TPM, success is virtually guaranteed.

Once goals have been set, each employee must understand and identify with them and develop small group activities in the workplace to ensure their achievement. In TPM, small groups set their own goals based on the overall company goals.

Figure 1-5 is an example of a companywide TPM policy and the goals associated with it. In 1981, Tokai Rubber Industries approached zero breakdowns and won the PM Prize. The firm was striving not simply to win, but to achieve its company goals — through TPM.

Developing a TPM Master Plan

To implement TPM over the course of three years, for example, one needs a master plan. Once this plan has been developed, it serves as a schedule for TPM that can be broken down into distinct stages. Figure 1-6 shows the TPM master plan used at Tokai Rubber Industries, which structures the development of TPM around three essential improvement goals:

- Autonomous maintenance through small group activities in the production department
- Refinement of preventive maintenance by the maintenance department and maintainability improvements to prevent equipment deterioration
- Reduction of startup failures through the application of maintenance prevention techniques at the equipment design stage

Basic TPM Policy
- Maximize equipment effectiveness through TPM
- Improve quality, increase safety, and reduce costs
- Raise morale

Goal
- Win the PM prize by August, 1981
- Show overall improvement in company conditions

Major Activities
1. Increase productivity by reducing breakdowns; assure safety and delivery
2. Increase productivity and reduce inventory by reducing setup and adjustment time
3. Improve quality and reduce losses by constantly monitoring equipment and die precision
4. Reduce cost by conserving materials and energy
5. Increase existing equipment effectiveness and prolong life span through improvement activities
6. Educate employees to raise morale and keep pace with increasingly sophisticated equipment

Figure 1-5. Example of Basic TPM Policy (Tokai Rubber Industries)

TPM Promotional Structure

TPM goals are implemented through "bottom-up" small group activities on the shop floor that must be constantly promoted by management and workers (Figure 1-7). TPM's organizational structure, similar to Rensis Lickert's "participative

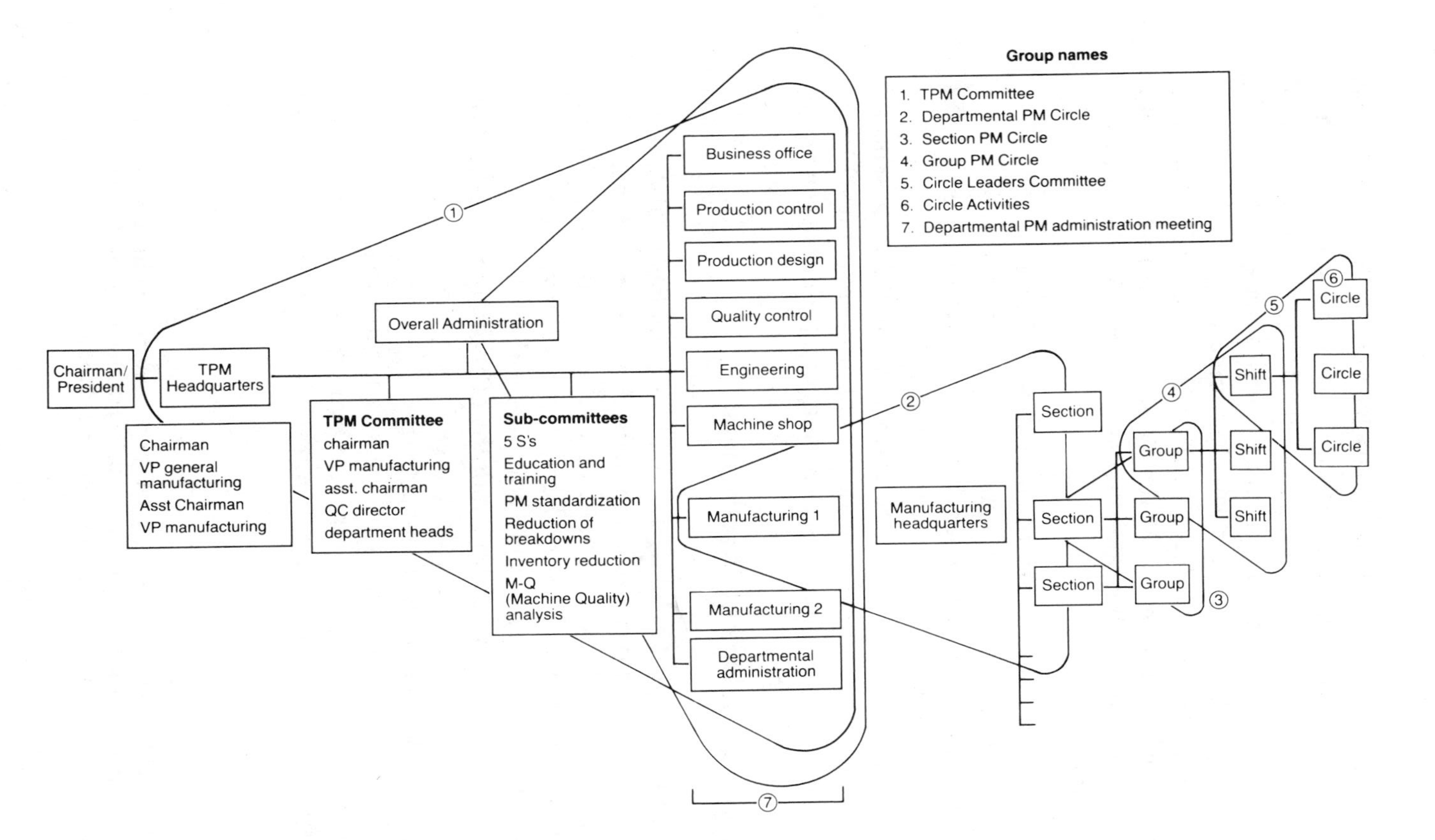

Figure 1-8. TPM Promotional Structure at Aisan Industries

however, must be determined for each company individually. The program must be adjusted to fit individual requirements, since the types of industry, production methods, equipment conditions, special needs and problems, techniques, and levels of maintenance vary from company to company.*

There are five interdependent goals, representing the minimum requirements for TPM development (Figure 1-9). Briefly summarized here, they are discussed in greater detail in subsequent chapters:

- Improving equipment effectiveness (2 and 3)
- Autonomous maintenance by operators (4 and 8)
- A planned maintenance program administered by the maintenance department (3 and 5)
- Training to improve operation and maintenance skills (7)
- An early equipment management program to prevent problems occurring during new plant or equipment startup (6)

Improving Equipment Effectiveness

Model projects help demonstrate the potential of TPM during the initial stages of its development. Several project teams are formed, consisting of engineering and maintenance staffs and production line supervisors. Equipment suffering from chronic losses is selected, preferably equipment that can be improved significantly in around three months of thorough investigation and analysis. Each project team targets one of the six big losses as a focus for their improvement activity.

When positive results are achieved, the project can be expanded to other similar equipment, with project team members pursuing further small group improvement activities in their own areas.

* When the JIPM is consulted by a Japanese company wishing to introduce TPM, it performs a thorough investigation of the industry involved and offers a program appropriate for the company's needs and problems.

Autonomous Maintenance by Operators

Autonomous maintenance by operators is one of the most distinctive features of TPM. The longer a company has functioned according to the concept of the division of labor, however, the more its employees will be convinced that the work of operators and maintenance workers should be strictly separated.

A company's established patterns of thinking and atmosphere cannot be changed overnight. Changing corporate culture takes two to three years, depending on company size. Operators accustomed to thinking "I operate — you fix" will have difficulty learning "I'm responsible for my own equipment." All employees must agree that operators are responsible for the maintenance of their own equipment; in addition, the operators themselves must be trained in the skills necessary for autonomous maintenance.

In many factories, operators already check and lubricate their own equipment, but they often do so grudgingly, without enthusiasm or understanding. For example, a worker may fill in the daily inspection sheet several days in advance or forget to refill the oil dispenser. Such incomplete care results in abrasion, wear, vibration, dirt, and deterioration, and it may lead to breakdowns and quality defects in process.

In Japan, the basic principles of industrial housekeeping are known as the Five S's: *seiri* (organization), *seiton* (tidiness), *seiso* (purity), *seiketsu* (cleanliness), and *shitsuke* (discipline).*

Often these principles are implemented only on a superficial level (*e.g.*, painting floors and equipment), while actual maintenance of equipment is inadequate (*e.g.*, neglecting the interior of the equipment, such as revolving parts). This superficiality is avoided in TPM's autonomous maintenance by breaking down training and practice into seven steps (one involves applying

* While these terms have very general meanings in translation, in actual practice each term refers to a specific principle or a set of established rules of organization and housekeeping. These specific meanings vary widely from company to company. The number of S's promoted by a company can also vary from as few as three to as many as seven.

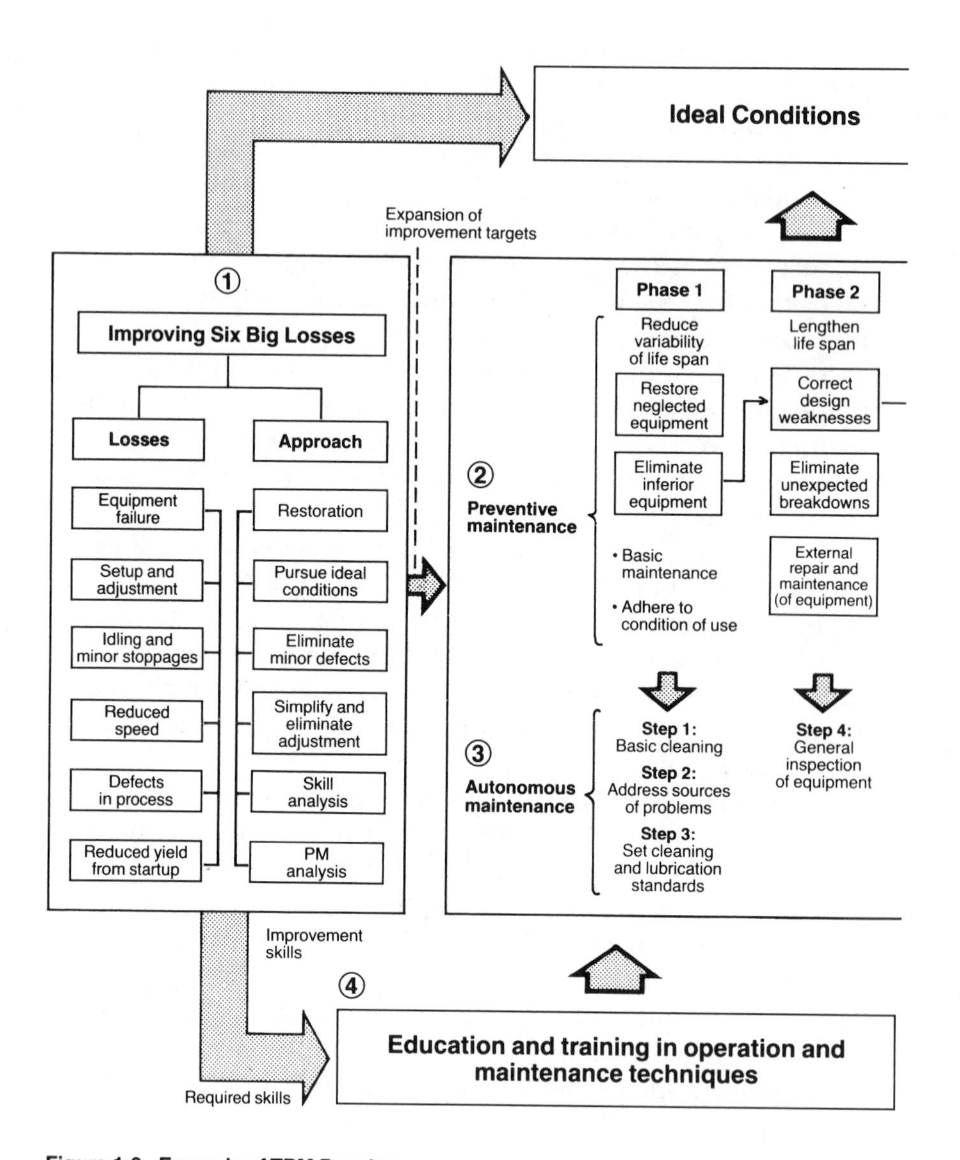

Figure 1-9. Example of TPM Development

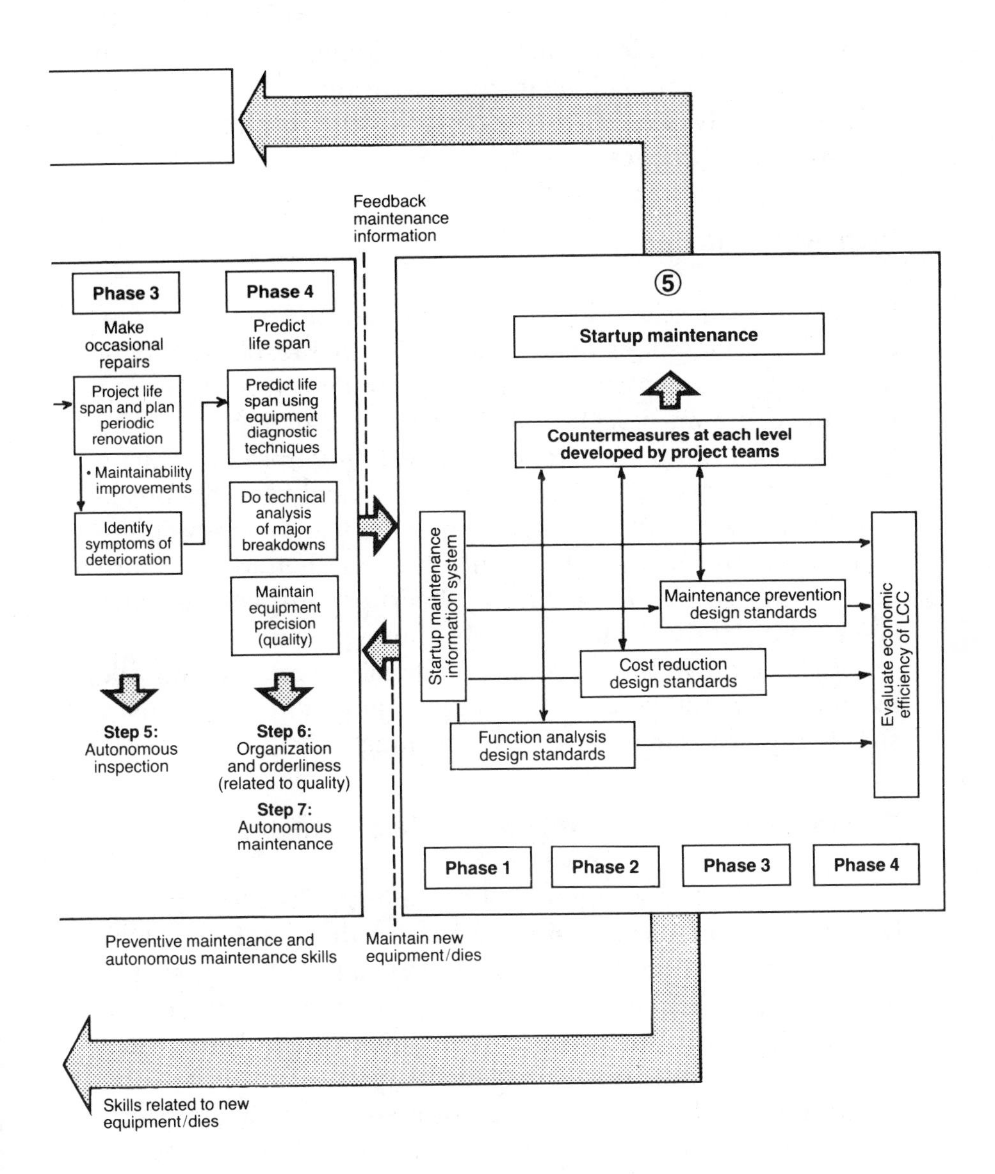
Feedback maintenance information
Phase 3
Make occasional repairs
Project life span and plan periodic renovation
• Maintainability improvements
Identify symptoms of deterioration
Phase 4
Predict life span
Predict life span using equipment diagnostic techniques
Do technical analysis of major breakdowns
Maintain equipment precision (quality)
Step 5: Autonomous inspection
Step 6: Organization and orderliness (related to quality)
Step 7: Autonomous maintenance
Preventive maintenance and autonomous maintenance skills
Maintain new equipment/dies
⑤
Startup maintenance
Countermeasures at each level developed by project teams
Startup maintenance information system
Maintenance prevention design standards
Cost reduction design standards
Function analysis design standards
Evaluate economic efficiency of LCC
Phase 1
Phase 2
Phase 3
Phase 4
Skills related to new equipment/dies

the principles of the Five S's). The tasks involved in each step must be thoroughly mastered before operators are introduced to the next. For example, at step 1, *initial cleaning,* operators learn that thorough cleaning is an inspection process. They learn a set of rigorous daily cleaning checkpoints as well as basic lubricating and bolting techniques.

Planned Maintenance

Planned or scheduled maintenance must function in tandem with autonomous maintenance. The first responsibility of the maintenance department is to deal with demands originating from the operators quickly and effectively. Maintenance personnel must also eliminate deterioration resulting from inadequate lubrication and cleaning. Then, they must analyze every breakdown to reveal weaknesses in the equipment and modify equipment to improve its maintainability and lengthen its life. Once maintenance costs are reduced, checkups, inspections, and equipment standards should also be thoroughly reviewed.

To keep down the cost of planned maintenance, diagnostic techniques should be used to monitor equipment conditions; a shift to predictive maintenance is encouraged.

Training to Improve Operating and Maintenance Skills

Some people may argue that operating skills and expertise become superfluous with increased automation. Unfortunately, while unmanned production may be possible, fully automated maintenance is unlikely. The skills of operators and maintenance personnel must be improved if autonomous maintenance, predictive maintenance, and maintainability improvement — the basic methods of TPM — are to be successful. Training in operation and maintenance skills is vital. To implement TPM, a company must be willing to invest in training its employees in the use of their equipment.

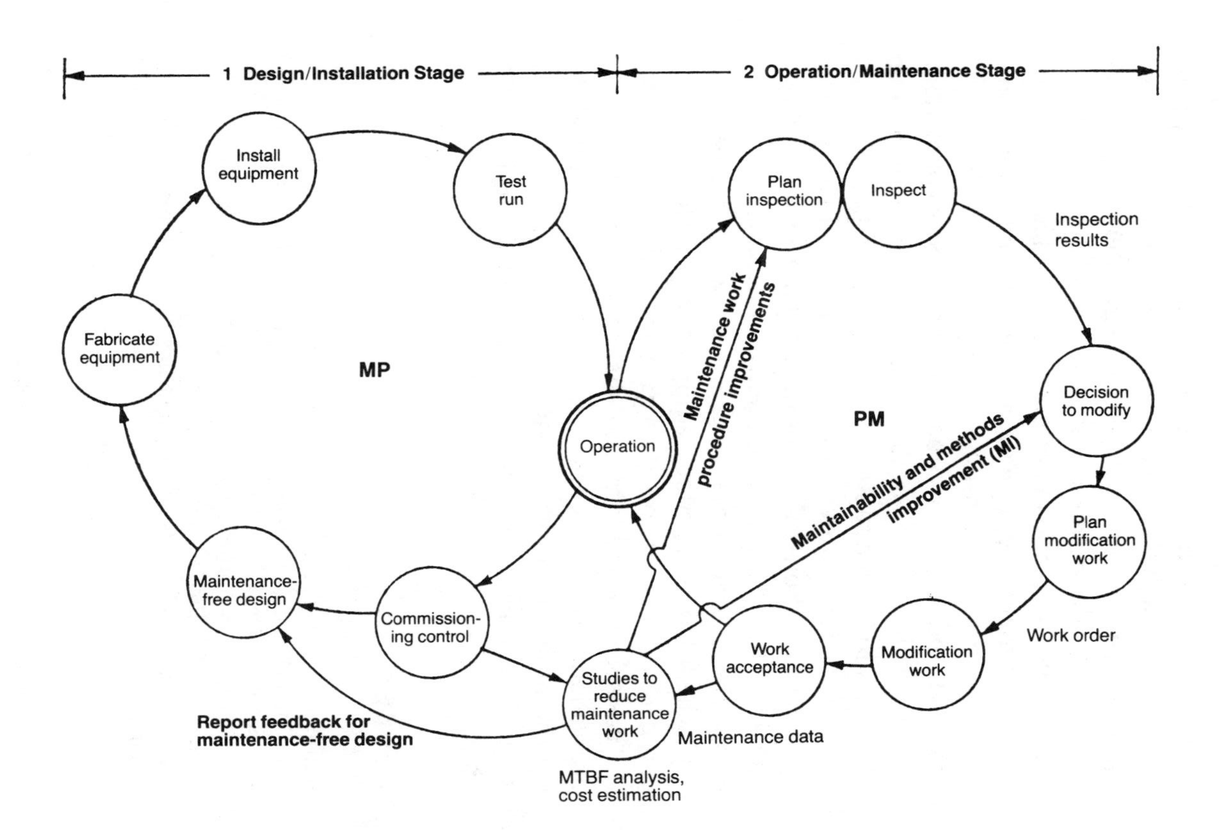

Figure 1-10. Model for Maintenance-Free Equipment Design

Early Equipment Management

Ideally, equipment should not require maintenance. A system that helps us approach this ideal is extremely valuable. Figure 1-10 represents a model system to promote maintenance-free design. On the left side are the design and installation stages; on the right, the operation and maintenance stages. The life cycle of a piece of equipment begins with a design aimed at reducing maintenance as much as possible. The equipment is then fabricated, installed, and tested before being put into normal operation. Once the early failure period of the equipment is over, operating data is fed back to the maintenance-free design stage. This data can be used to design maintenance-free equipment in the future. Maintenance prevention (MP) is the object of the design-installation cycle, including startup equipment maintenance.

During the operation-maintenance stage, on the basis of regularly scheduled inspection, equipment is restored, modified, or replaced. Maintenance data gathered in this process provides the basis for research on maintenance prevention.

Information feeds back for three types of improvement: (1) to improve the maintainability of equipment now in use, (2) to improve maintenance work and systems, and (3) to facilitate maintenance-free design in new equipment.

REFERENCES

Eighth Terotechnology Survey Mission to Europe (JIPM) (May 1980).

Kuroda, M. "Achieving zero defects through employee participation (3)" (in Japanese). *Plant Engineer* 12 (November 1980) 50.

Ohno, Taiichi. *Toyota Production System: Beyond Large-Scale Production*. Cambridge: Productivity Press, 1988.

Okuda, S. "Achieving zero defects through employee participation (1)" (in Japanese). *Plant Engineer* 12 (September 1980) 43.

2

Equipment Effectiveness, Chronic Losses, and Other TPM Improvement Concepts

Equipment effectiveness is a measure of the value added to production through equipment.*

TPM maximizes equipment effectiveness through two types of activity:

- *quantitative*: increasing the equipment's total availability and improving its productivity within a given period of operating time
- *qualitative*: reducing the number of defective products, stabilizing, and improving quality

The goal of TPM is to increase equipment effectiveness so each piece of equipment can be operated to its full potential and maintained at that level. Human workers and machinery should both function steadily under optimal conditions with zero breakdowns and zero defects. Although approaching zero is difficult,

---

* Simply stated, *added value* is the difference between the sales revenue and the cost of resources (material and labor) used to produce a product. The value added to a product by equipment is significantly reduced by waste and the six major equipment-related losses. It is increased as equipment availability and productivity go up and defects in process and rework go down.

believing that zero defects can be achieved is an important prerequisite for the success of TPM.

The following sections address some of the difficulties inherent in maximizing equipment effectiveness.

SIX BIG LOSSES LIMIT EQUIPMENT EFFECTIVENESS

Equipment effectiveness is limited by the following six types of losses.

Breakdown Losses

Two types of loss are caused by breakdowns: *time losses,* when productivity is reduced, and *quantity losses,* caused by defective products.

Sporadic breakdowns — sudden, dramatic, or unexpected equipment failures — are usually obvious and easy to correct. Frequent or chronic minor breakdowns, on the other hand, are often ignored or neglected after repeated unsuccessful attempts to cure them. Because sporadic breakdowns account for a large percentage of total losses, factory personnel invest a great deal of time and effort searching for ways to avoid them. Eliminating them is extremely difficult, however. Typically, studies to increase equipment reliability must be conducted and ways found to minimize the time needed to correct problems when they occur.

To maximize equipment effectiveness, however, *all* breakdowns must be reduced to zero. This is actually possible without much effort or investment — although some investment may be necessary in the beginning. First, however, the conventional philosophy of breakdown maintenance — the belief that breakdowns are inevitable — must be changed.

Setup and Adjustment Losses

Losses during setup and adjustment result from downtime and defective products that occur when production of one item ends and the equipment is adjusted to meet the requirements of another item. Many companies are now working to achieve single-minute setups (under 10 minutes). Working from an industrial engineering perspective, setup time can be reduced considerably by making a clear distinction between *internal* setup time (operations that must be performed while the machine is down) and *external* setup time (operations that can be performed while the machine is still running) and by reducing internal setup time. (*See* Chapter 3, pp. 112-119.)

Idling and Minor Stoppage Losses

A minor stoppage occurs when production is interrupted by a temporary malfunction or when a machine is idling. For example, some workpieces might block the top of a chute, causing the equipment to idle; at other times sensors, alerted by the production of defective products, shut down the equipment. These types of temporary stoppage clearly differ from a breakdown. Normal production is restored by simply removing the obstructing workpieces and resetting the equipment.

Small problems like this often have a dramatic effect on equipment effectiveness, however, typically when robots, automatic assemblers, conveyors, and so on, are involved. Minor stoppages and idling, while easily remedied, are also easily overlooked because they are often difficult to quantify. For this reason, the extent to which minor stoppages hinder equipment effectiveness often remains unclear.

Zero minor stoppages is an essential condition for unmanned production. If minor stoppages are to be reduced, operating conditions must be closely observed and all slight defects must be eliminated. (*See* Chapter 3, pp. 133-149.)

Reduced Speed Losses

Reduced speed losses refer to the difference between equipment design speed and actual operating speed. Speed losses are typically overlooked in equipment operation, although they constitute a large obstacle to equipment effectiveness and should be studied carefully. The goal must be to eliminate the gap between design speed and actual speed.

Equipment may be run at less than ideal or design speed for a variety of reasons: mechanical problems and defective quality, a history of past problems, or fear of abusing or overtaxing the equipment. Often, the optimal speed is simply not known. On the other hand, deliberately increasing the operating speed actually contributes to problem-solving by revealing latent defects in equipment conditions. (*See* Chapter 3, pp. 150-153.)

Quality Defects and Rework

Quality defects in process and rework are losses in quality caused by malfunctioning production equipment. In general, sporadic defects are easily and promptly corrected by returning equipment conditions to normal. These defects include sudden increases in the quantity of defect, or other dramatic phenomena. The causes of chronic defects, on the other hand, are difficult to identify. Ad hoc measures to restore the status quo rarely solve the problem, and the conditions underlying the defects may be ignored or neglected. Defects requiring rework should also be counted as chronic losses.

Reducing chronic defects, like reducing chronic breakdowns, requires thorough investigation and innovative remedial action. The conditions surrounding and causing the defect must be assessed and control limits evaluated. Complete elimination of defects is, as always, the main goal. (*See* Chapter 3, pp. 153-164.)

Startup Losses

Startup losses are yield losses that occur during the early stages of production — from machine startup to stabilization. The volume of losses varies with the degree of stability of processing conditions; maintenance level of equipment, jigs, and dies; operators' technical skills; and so on. In practice, the volume is surprisingly high. Such losses are latent, and the possibility of eliminating them is often obscured by uncritical acceptance of their inevitability.

Table 2-1 sets out the improvement goals for the preceding losses. Table 2-2 illustrates the possible levels of overall equipment effectiveness.

Type of Loss	Goal	Explanation
1. Breakdown losses	0	Reduce to zero for all equipment
2. Setup and adjustment losses	minimize	Reduce setups to less than ten minutes
3. Speed losses	0	Bring actual operation speed up to design speed; then make improvements to surpass design speed
4. Idling and minor stoppage losses	0	Reduce to zero for all equipment
5. Quality defect and rework losses	0	Extremely slight occurrences acceptable (*e.g.*, 100-30 ppm)
6. Startup (yield) losses	minimize	

Table 2-1. Improvement Goals for Chronic Losses

MEASURING EQUIPMENT EFFECTIVENESS

Effectiveness can be measured using the formula:

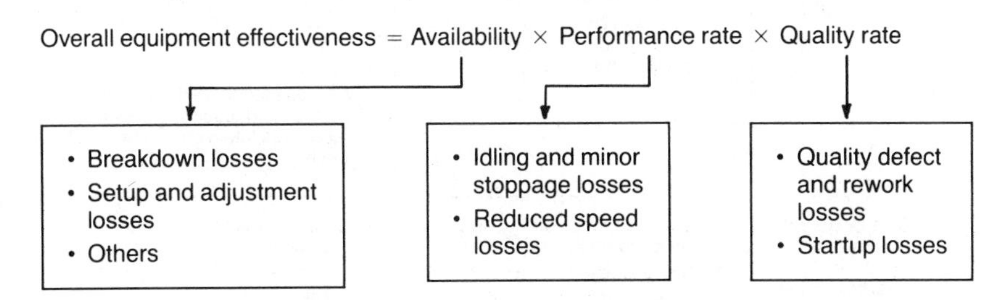

TPM is not limited to dealing with breakdowns; rather it raises the level of total equipment effectiveness by improving all related factors:

- *availability* (operating rate): improved by eliminating breakdowns, set-up/adjustment losses, and other stoppage losses

Assessment	Level 1	Level 2
1. Breakdown losses	1. Combination of sporadic and chronic breakdowns 2. BM>PM 3. Significant breakdown losses 4. Autonomous operator maintenance not organized 5. Unstable life spans 6. Equipment weaknesses are unrecognized	1. Sporadic breakdowns 2. PM≈BM 3. Breakdown losses still significant 4. Autonomous maintenance being organized 5. Parts life spans estimated 6. Equipment weaknesses well-acknowledged 7. MI (maintainability improvement) applied on above points
2. Setup and adjustment losses	1. No control — *laissez-faire* production by operators 2. Work procedures disorganized and setup/adjustment time varies widely	1. Work procedures organized (*e.g.*, internal and external setup distinguished 2. Setup/adjustment time unstable 3. Problems to be improved identified
3. Speed losses	1. Equipment specifications not well understood 2. No speed standards (by product and machinery)	1. Problems related to speed losses analyzed • Mechanical problems • Quality problems 2. Tentative speed standards set and maintained by product 3. Speeds vary slightly
4. Idling and minor stoppage losses	1. Losses from minor stoppages unrecognized 2. Unstable operating conditions due to fluctuation in frequency and location of losses	1. Minor stoppage losses analyzed quantitatively • Frequency and location of occurence • Volume lost 2. Losses categorized and outbreak mechanism analyzed; preventive measures taken on trial-and-error basis
5. Quality defect and rework losses (startup losses included)	1. Chronic quality defect problems are neglected 2. Many successful remedial actions have been taken	1. Chronic quality problems quantified • Details of defect, frequency • Volume lost 2. Losses categorized and outbreak mechanism explained; preventive measures taken on trial-and-error basis

Table 2-2. Assessment of Overall Equipment Effectiveness

Level 3	Level 4
1. Time-based maintenance established 2. PM>BM 3. Breakdown losses less than 1% 4. Autonomous maintenance activities well-established 5. Parts life spans lengthened	1. Condition-based maintenance established 2. PM 3. Breakdown losses 0.1% – 0 4. Autonomous maintenance activities stable and refined 5. Parts life spans predicted 6. Reliable and maintainable design developed
1. Moving internal setup operations into external setup time 2. Adjustment mechanisms identified and well understood	1. Setup time less than 10 minutes 2. Immediate product changeover by eliminating adjustments
1. Necessary improvements being implemented 2. • Speed is set by the product • Cause-and-effect relationship predicted between the problem and the precision of the equipment, jigs, and tools 3. Small speed losses	1. Operation speed increased to design speed or beyond through equipment improvements 2. Speed standards set and maintained by product (final standards) 3. Zero speed losses
1. All causes of minor stoppages analyzed, all solutions implemented, conditions favorable	1. Zero minor stoppages (unmanned operation possible)
1. All causes of chronic quality defects analyzed, all solutions implemented, conditions favorable 2. Automatic in-process detection of defects under study	1. Quality losses = 0.1% – 0

- *performance*: improved by eliminating speed losses, minor stoppages, and idling
- *quality* (rate of quality products): improved by eliminating quality defects in process and during startup

The operating, performance, and quality rates can be determined in each work center, but the importance of each factor varies according to the characteristics of the product, equipment, and production systems involved. For example, if adjustments and breakdowns are high, the operating rate will be low, and if many minor stoppages occur, the performance rate will be low. A high level of equipment effectiveness can be achieved only when all three rates are high.

The following principles must be applied when improving equipment effectiveness:

- Make detailed, accurate measurements.
- Set firm priorities.
- Establish clear directions or goals.

Calculating the Operating, Performance, and Quality Rates

Figure 2-1 illustrates the relationship of the six major losses to the three rates used for calculating equipment effectiveness.

Loading time refers to the net availability of equipment during a given period, such as a day or month. In other words, it is the total time available for operation minus planned or necessary downtime such as breaks in production schedule, precautionary resting times, and daily shop floor meetings. *Operating time* is the loading time minus the time the machine is down due to breakdowns, setup and adjustments, retooling, and other stoppages. In other words, it refers to the time during which the equipment is actually in operation.

Net operating time is the time the equipment is operated at a stable or constant speed. Time losses due to minor stoppages and operating at reduced speed (often estimated) are subtracted from the operating time to determine the net operating time.

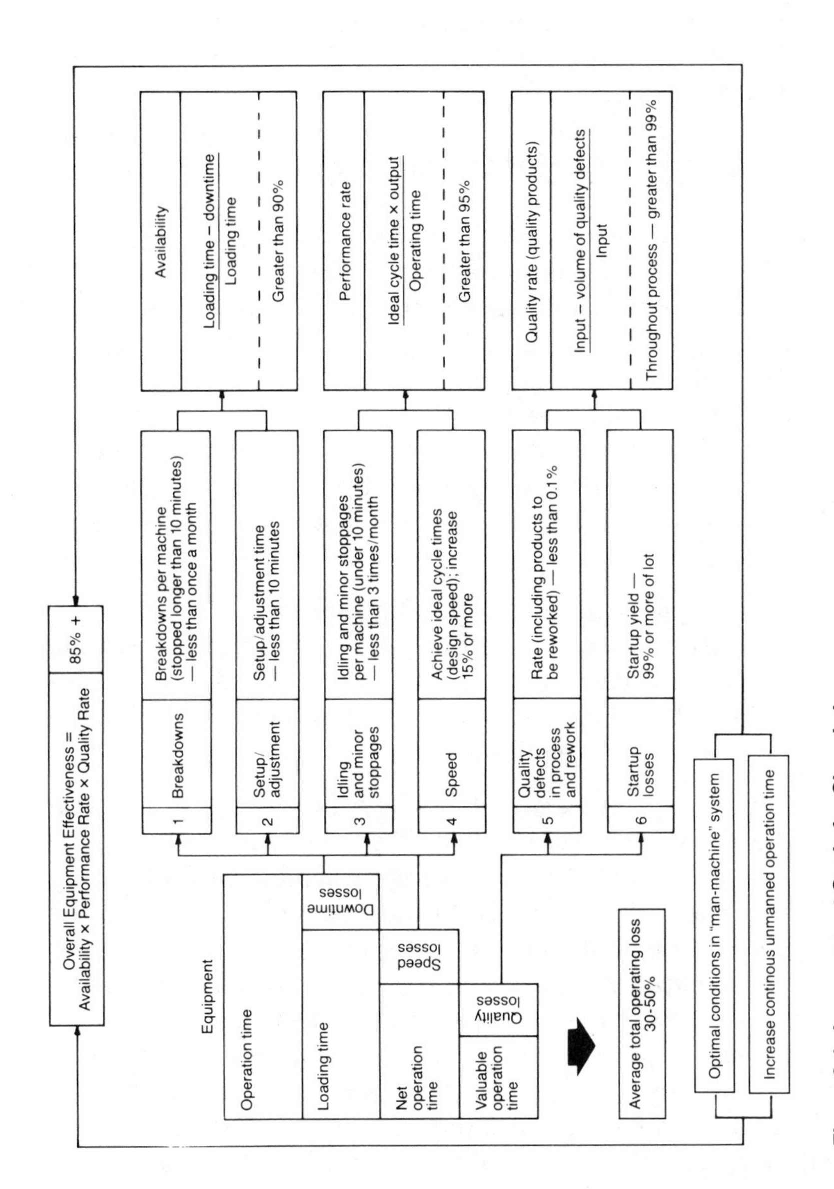

Figure 2-1. Improvement Goals for Chronic Losses

Valuable operating time is the net operating time minus the estimated time required to rework defective products. It is the time during which acceptable products are manufactured.

Availability, or the *operating rate,* is the ratio of the loading time to the net operating time.

The *performance rate* is based on the operating speed rate and the net operating time. The *operating speed rate* is the ratio of the ideal or design cycle time of the equipment to the actual cycle time, which reflects reduced speed losses.

Although the design cycle time is often used in this ratio, in some cases the operating speed must be lower than the design speed for quality reasons, when design defects produce substandard products, for example, or when operating at design speed results in equipment problems. In such cases, performance should be calculated using the lower cycle time. Obviously, speed must be determined on a case-by-case basis. Depending on equipment conditions, one of the following methods should be used:

- cycle time determined by design speed
- cycle time based on current optimal conditions (cycle time changes according to product)
- highest cycle time achieved or cycle time estimated on the basis of similar equipment

The net operating rate depends on maintaining a certain speed over a given period of time. Thus, losses from minor stoppages, as well as those from correcting small problems and making adjustments, must be taken into consideration. Ultimately, the actual speed (regardless of the design or standard speed) is irrelevant. Equipment can certainly be operated at a lower speed, as long as stable, long-term operation can be maintained.

Overall equipment effectiveness is the product of the operating rate (availability), the performance rate, and the quality rate. This measurement combines the current availability and speed of the equipment with its quality rate. It reflects the overall capability of the plant.

- Availability (operating rate) $= \dfrac{\text{loading time} - \text{downtime}}{\text{loading time}}$

- Performance rate $= \dfrac{\text{output} \times \text{actual cycle time}}{\text{loading time} - \text{downtime}} \times \dfrac{\text{ideal cycle time}}{\text{actual cycle time}}$

  Net operating rate reflects losses resulting from minor stoppages

  Operating speed rate reflects reduced speed losses

- Quality rate $= \dfrac{\text{number of good products}}{\text{input}}$

  Number of good products =
  Input − (startup defects + process defects + trial products)

- Overall equipment effectiveness =
  Availability × Performance × Quality

Example

Working hours per day: 60 min. × 8 hours = 480 minutes
Loading time per day: 460 minutes
Downtime per day: 60 minutes
Operating time per day: 400 minutes
Output per day: 400 products
Types of downtime:
  Setup — 20 minutes
  Breakdowns — 20 minutes
  Adjustments — 20 minutes
Defects — 2%
Availability (operating rate) = 400 ÷ 460 × 100 = 87%
Ideal cycle time: 0.5 minutes per product
Actual cycle time: 0.8 minutes per product
Operating speed rate = 0.5 ÷ 0.8 × 100 = 62.5%
Net operating rate = 400 pcs. × 0.8 ÷ 400 minutes × 100 = 80%
(100 − net operating rate) reflects losses caused by minor stoppages
Performance rate = 0.625 × 0.800 × 100 = 50%
Quality rate = 98%
Overall equipment effectiveness = 0.87 × 0.5 × 0.98 × 100 = 42.6%

Levels and Targeted Goals for Overall Effectiveness

Levels for overall effectiveness differ depending on the industry, equipment features, and production systems involved. Equipment effectiveness averaged from 40 to 60 percent at the companies investigated by JIPM (Table 2-3). This standard can be raised to as much as 85 to 95 percent, however, through various TPM improvement activities. Table 2-4 (see pages 40-41) shows the actual conditions and figures at one participating firm.

Criteria	Automated machinery	Automatic assemblers	Automatic packers
1. Overall effectiveness	51.3-78.4	38.0-80.7	72.0%
2. Availability	95-98	95	90
3. Performance	54-80	40-85	80
‖			
Operating speed rate	90-100	100	100
×			
Net operating rate	60-80	40-85	80
Remarks	**20-40% of losses due to idling and minor stoppages**	**15-60% of losses due to idling and minor stoppages**	**20% of losses due to idling and minor stoppages**

Table 2-3. Sample Overall Effectiveness Conditions

CHRONIC LOSSES AND HIDDEN DEFECTS

Chronic losses are caused by hidden defects in machinery, equipment, and methods. If fundamental conditions in the manufacturing environment are to improve, chronic losses and hidden defects must be completely eliminated. The remainder of this chapter describes the nature of chronic losses and outlines a methodology for detecting and eliminating hidden defects.

Until now, maintenance has typically addressed problems that can be characterized as sporadic — infrequent or unusual events that cause a sudden breakdown or obvious loss of quality. Chronic losses, by contrast, are subtle and much more difficult to detect. Their causes can be exposed and eliminated, however, by changing the approach to factory maintenance.

Chronic Losses and Sporadic Losses

The term *chronic* usually refers to a phenomenon that occurs repeatedly within a certain range of distribution. Sudden outbreaks that go beyond this range are referred to as *sporadic*. These sporadic outbreaks can take the form of either an increase in the quantity of a particular phenomenon or a completely different phenomenon.

The remedy for sporadic losses is *restoration*, since they are usually triggered by changes in conditions (*e.g.*, equipment, jigs and tools, work methods, or operating conditions). They can be corrected by taking actions to restore conditions to normal levels (Figure 2-2).

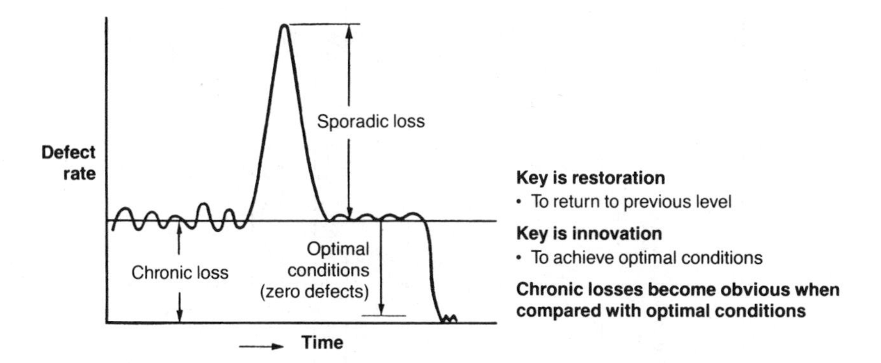

Figure 2-2. Sporadic and Chronic Losses

The key to chronic losses, on the other hand, is *innovation*. Chronic problems tend to resist traditional remedies because their roots are hidden in the structure of the equipment and the methods used. They arise from conditions that have come to be perceived as normal, so restorative action or quick-fix remedies

Process	A	B	C	D	E	F	T	G		H
	Working hours (h)	Planned downtime (h)	Loading time (h) A − B	Downtime loss (h)	Operating time C − D	Actual processing time (h) J × G	Availability E/C × 100 (%)	Quantity processed: Output (number of quality products) G_1	Quantity processed: Total (including losses and rework) G	Quality rate G_1/G
Internal Finish	17.75	2.3	15.45	—	15.45	10.6	100	4,000	4,018	99.6
	18.58	0.6	17.98	—	17.98	12.8	100	4,800	4,818	99.6
	16.83	0.7	16.13	—	16.13	11.65	100	4,400	4,418	99.6
Total	(59.66)	(3.7)	(55.96)	—	(56.06)	(39.05)	(100)	(14,800)	(14,863)	(99.55)
External Finish	6.5	0.1	6.4	—	6.4	4.62	100	1,187	1,190	99.7
	17.75	0.7	17.5	0.5	16.55	13.8	97.1	3,513	3,542	99.2
	17.75	0.8	16.95	—	16.95	12.86	100	3,297	3,309	99.6
	16.0	1.1	14.9	—	14.9	12.25	100	3,030	3,150	96.2
Total	(58.0)	(1.7)	(56.3)	(0.5)	(54.8)	(43.62)	(99.27)	(11,027)	(11,191)	(98.67)
External Finish	6.5	0.1	6.4	0.2	6.2	3.17	96.9	1,187	1,202	98.8
	17.75	0.7	17.05	—	17.05	9.48	100	3,513	3,593	99.8
	17.75	0.8	16.95	—	16.95	8.95	100	3,297	3,392	97.2
	16.0	1.1	14.9	—	14.9	8.1	100	3,030	3,060	99.0
Total	(58.0)	(1.7)	(56.3)	(0.2)	(55.1)	(29.7)	(99.22)	(11,027)	(11,247)	(98.2)

Process	I	J	L	M	N	Number of minor stoppages	Number of exchanges of diamond grindstone	Overall equipment effectiveness
	Ideal cycle time (s) M/c	Actual cycle time (s) M/c	Performance rate M × N (%)	Operation speed rate I/J (%)	Net operating rate F/E (%)			T × L × H
Internal Finish	9.5	9.8	66.4	96.9	68.6	2	2	66.1
	9.5	9.8	68.9	96.9	71.2	4	2	68.6
	9.5	9.8	69.9	96.9	72.2	—	2	69.6
Total	(9.5)	(9.78)	(66.9)	(97.07)	(69.7)	(8)	(7)	66.5
External Finish	14	16.6	60.6	84.3	72.2	1	3	60.4
	14	16.65	70.3	84.1	83.4	—	8	67.6
	14	16.45	64.5	85.1	75.9	—	6	64.2
	14	16.15	71.2	86.7	82.2	—	6	68.1
Total	(14)	(16.45)	(66.3)	(85.05)	(78.42)	(1)	(23)	65.6
External Finish	9.5	10.4	46.4	91.3	51.1	—	1	43.6
	9.5	10.5	50.0	90.1	55.6	—	4	49.9
	9.5	10.5	47.5	90.1	52.8	—	3	45.5
	9.5	10.4	49.6	91.3	54.4	—	3	45.1
Total	(9.5)	(10.45)	(48.4)	(90.7)	(53.47)	—	(11)	48.4

Table 2-4. Calculating Overall Equipment Effectiveness

have no effect. Chronic losses can be reduced only through *break-through* thinking — abandoning conventional tactics for new and creative methods that pursue and eliminate the hidden causes. According to J. M. Juran:

> A sporadic condition is a sudden adverse change in the status quo, requiring remedy through *restoring* the status quo (*e.g.*, changing a worn cutting tool). A chronic condition is a longstanding adverse situation, requiring remedy through *changing* the status quo (*e.g.*, revising a set of unrealistic tolerances).*

Differences Between Chronic and Sporadic Losses

Chronic and sporadic losses differ in other ways as well:

Latency. Sporadic losses are conspicuous, because they create conditions that differ considerably from routine operating conditions. Chronic problems, by contrast, tend to remain hidden. Typically, they are hard to measure, easily overlooked, or ignored because of cynicism or preconceived notions about their causes.

Often, chronic losses can often be brought out into the open only by comparing present conditions with theoretical or optimal levels. For example, a piece of equipment may be designed to run at 250 spm (strokes per minute). If it is currently operating at only 200 spm, the speed loss of 50 spm will remain hidden if clear standards for equipment capacity have not been specified. Only when the current speed is compared to the standard speed does the loss become evident (Table 2-5).

Here is another example. Setup for a particular machine currently takes an hour. If technical and operational improvements can reduce the setup time to 30 minutes, then a latent 30-minute loss is exposed.

Causation. Cause-and-effect relationships are relatively simple to trace in the case of sporadic losses; in chronic losses, on

* J. M. Juran and F. M. Gryna, Jr., *Quality Planning and Analysis: From Product Development through Usage* (New York: McGraw-Hill Book Co., 1970), 9.

Loss	Obvious	Hidden
1. Sporadic breakdowns	×	
Chronic breakdowns		×
2. Setup and adjustment	×	×
3. Idling and minor stoppage		×
4. Speed		×
5. Sporadic quality defects	×	
Chronic quality defects		×

Table 2-5. Characteristics of Chronic Loss

the other hand, they are often unclear. A single cause is rare — a combination of causes tends to be the rule.

Types of remedial action. Measures against sporadic problems are not very difficult to develop because, as explained above, their causes are easily determined. Since the causes of chronic losses are often complex, this type of loss can remain unsolved even after numerous remedies have been attempted.

Economic impact. A single sporadic problem can be very costly compared to a single occurrence of a chronic loss. The cumulative effect and cost of these smaller losses is considerable, however, because of the frequency of their occurrence.

To summarize, sporadic problems are conspicuous and have clear causes, so appropriate actions are relatively easy to design. Chronic problems

- are usually latent
- result in negligible loss per incident
- occur frequently
- can be easily restored by operators
- rarely come to the attention of supervisors
- are difficult to quantify
- must be detected through comparison with optimal conditions

Most companies take some action to solve sporadic problems while leaving chronic problems essentially untouched.

Characteristics of Chronic Losses

To reduce chronic loss, its characteristics must be grasped fully. Causes cannot be assigned without a thorough understanding of the conditions surrounding a particular loss. The hasty conclusions based on inadequate investigation that can be observed in many companies must be avoided. The numerous, constantly changing causes of chronic problems produce a complex of phenomena that interact in different combinations in every occurrence. It is therefore more prudent to conclude that the cause is unknown. An effective remedy can often be developed for a single cause, but as long as other causes remain hidden, little or no improvement will result (Figure 2-3).

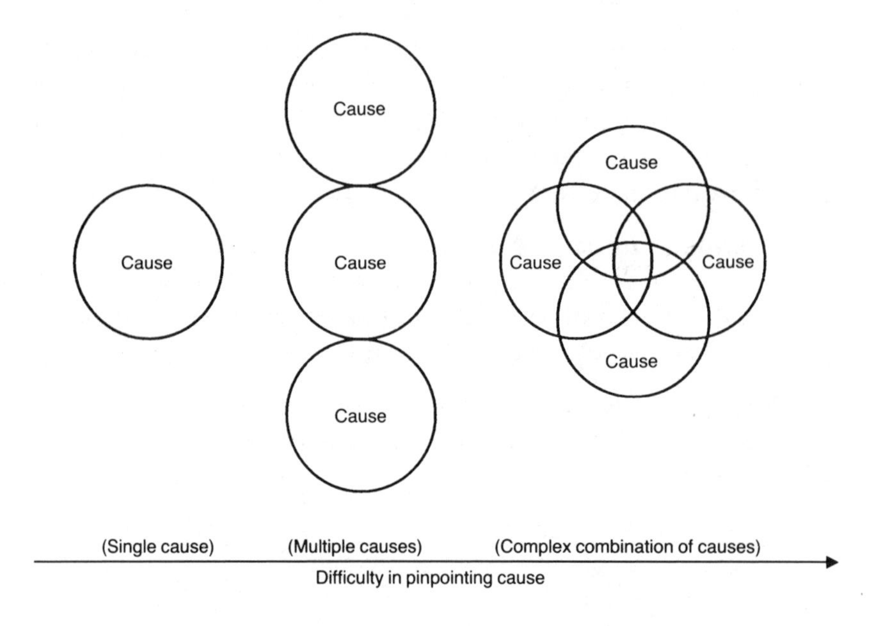

Figure 2-3. Causes of Chronic Losses

Do not plan remedial action before thoroughly investigating the peculiarities of a chronic loss. Moreover, *do not neglect or eliminate any possible causes.* Once all factors that might influence the loss have been identified, measures should be designed and implemented to control each of them.

How Chronic Losses Occur

Chronic losses occur under the following circumstances (Figure 2-4):

When the Loss is Recognized, But...

Remedial action has been unsuccessful. This is the most common situation. After various measures have been taken, results are unfavorable or insignificant, so the investigation is abandoned.

Remedial action cannot be taken. In this case, an effective countermeasure is obviously needed, but pressing production and delivery requirements leave no time to implement any radical or permanent solutions. Less effective temporary measures are taken instead, and the problem persists.

Remedial action is not taken. In this case, a chronic loss has been observed, but the extent of the problem is not recognized and the phenomena are ignored. Because the magnitude of such losses is not apparent, many companies underestimate their effect. The costs required to eliminate chronic losses mistakenly appear to outweigh the benefits, so the problems are often left untouched.

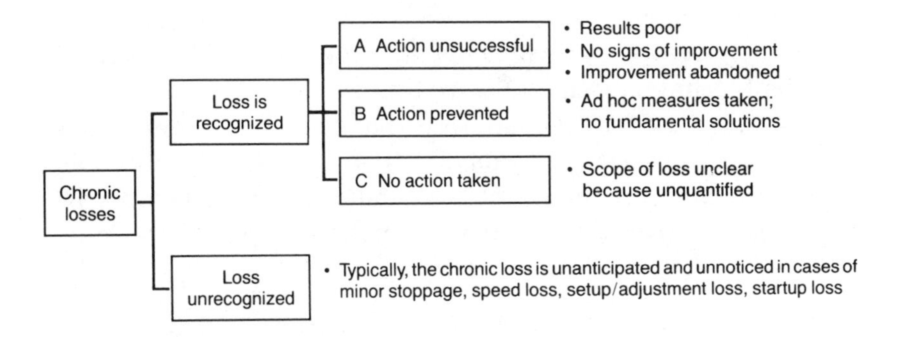

Figure 2-4. Background of Chronic Loss

When the Loss Is Unrecognized

Often a chronic loss goes unnoticed, especially when it is hidden in what appears to be a "normal" operation. Most involve minor stoppages (*e.g.*, temporary overload and blockage where operation resumes with removal of the obstacles), speed losses, rework, and startup losses.

Why Chronic Losses Are Neglected

The common reasons for neglecting chronic losses are summarized in Figure 2-5. They are the following:

The Cause Is Unknown

Discovering why a chronic loss occurs is difficult, because so many types of causes are possible: single causes, multiple causes, and complex causes, which occur in different combinations at different times.

Often engineers begin analyzing an equipment problem by taking an industrial engineering or quality control approach but fail to identify the cause. Next, they take preventive measures on a trial-and-error basis, but the problem still fails to improve. When this happens, the whole team becomes discouraged and the equipment is set aside because it "cannot be fixed," because "only new equipment will show an improvement," or because "current technology cannot solve this problem."

One reason for this pattern is that some engineers tend to adopt an overly narrow focus on certain details. Effective solutions, however, are the result of a detailed technical perspective balanced with overall understanding of the workplace (*e.g.*, equipment and production methods).

A Cause Is Known, but Action Taken Is Inappropriate

Because chronic losses are usually caused by a variety of interrelated factors, single actions provide only partial solutions.

Moreover, engineers often take inappropriate measures because they misunderstand the true nature of the loss after superficial observation. Or they reach hasty conclusions about its cause by relying on techniques that worked for them in the past.

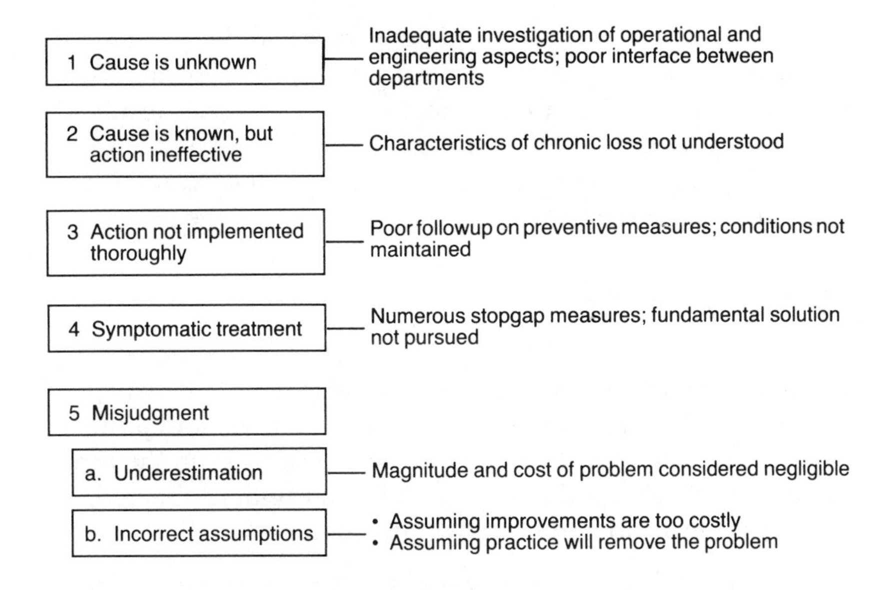

Figure 2-5. Why Chronic Losses Are Neglected

Remedial Action Is Incomplete

Sometimes an effective solution is identified but the results of implementation are disappointing because it was not carried out properly or thoroughly. In other words, simple procedural errors can trigger chronic problems. For example:

- The intended effect of the measure is not clearly communicated to the shop floor, so it is improperly implemented.
- Not all operators are equally well trained and knowledgeable; newcomers may inadvertently use inconsistent methods.
- The measures taken produce negative results because shop floor efforts are not thorough; lack of follow-up makes the problem worse.

All these problems are caused by lack of managerial planning and follow-up to check results.

Symptomatic Treatment of the Cause

Often, tight production and delivery schedules prevent implementation of long-term solutions and favor superficial, short-term remedies. Failure to treat root causes, however, results in the recurrence of similar problems, lower productivity, an increase in chronic losses, and production that cannot run smoothly. The earlier this vicious cycle is broken, the better.

Mistaken Approaches

Underestimation. Sporadic problems often produce dramatic losses that attract the attention of managers. The managers then provide prompt treatment, search for causes, and propose preventive measures. The scope of a chronic loss, on the other hand, is typically hard to detect and often underestimated, so managers are seldom informed of their occurrence.

Incorrect assumptions. Because the causes, magnitude, and frequency of chronic losses are so often misunderstood, it is easy to imagine that new equipment or changes in manufacturing processes, quality standards, or material are the only solutions. Thus, improving existing equipment and conditions appears costly and the potential results minimal. This is why chronic losses are often ignored.

Mistaking lack of training for lack of practice. In some cases, chronic losses seem inevitable because the work is always performed by a disproportionally large number of inexperienced workers. Managers incorrectly assume that practice and experience will eventually solve the problem. They do not analyze the skills of seasoned workers so that these skills can be easily transferred to newcomers. Where skill is a factor in the loss, new workers typically have not received adequate training in key operational steps, points to watch, and methods for checking results. Moreover, their work is often unchecked. This accounts for many chronic losses.

REDUCING AND ELIMINATING CHRONIC LOSSES

Chronic losses can be reduced and often eliminated by increasing equipment reliability, restoring the equipment to its original operating conditions, identifying and establishing optimal operating conditions, and eliminating small defects that are often overlooked.

Reliability

Equipment reliability is the probability that equipment, machinery, or systems will perform required functions satisfactorily under specific conditions within a certain period of time. It can also be thought of as the likelihood that problems (quality defects and breakdowns) *will not occur* over a given period. Low equipment reliability is the fundamental cause of chronic losses; it is also what makes them so difficult to eliminate. Low reliability leads to the incidence of quality defects and breakdowns, and these problems must be regarded as chronic when the intervals between occurrences are brief.

Intrinsic Reliability and Operational Reliability

Equipment reliability is based on two factors: intrinsic reliability and operational reliability. *Intrinsic reliability* is based on design and is determined during the design, fabrication, and installation stages. *Operational reliability* is determined by the user and is related to how and under what conditions the equipment is operated. *Total reliability* is the product of these two qualities. Reliability can be further subdivided:

Fabrication reliability. Faulty manufacturing and assembly of parts may result in poor dimensional accuracy, incorrect shapes of parts, and poor assembly.

Intrinsic reliability	Operational reliability
Design reliability Manufacturing reliability Installation reliability	Operation reliability Maintenance reliability

Installation reliability. Improper installation results in excessive vibration, equipment that is not level, and defective plumbing and wiring.

Design reliability. Faulty design includes jigs mismatched to workpiece shape, faulty mechanisms, short parts-life, parts misselection, poor instrumental detection systems, and so on.

Relatively few equipment defects are caused by poor design reliability, however. Most of them are related to operation, such as:

Operation and manipulation reliability. The following errors can reduce reliability in operation: manipulation errors, setup and adjustment errors, incorrect operation standards, and inconsistency in maintaining basic conditions (cleaning, lubrication, bolting).

Maintenance reliability. Equipment reliability is also reduced by maintenance errors, such as incorrect replacement of parts and incorrect assembly.

When breakdowns and quality defects occur, the source of low reliability must be investigated. Often it results from insufficient knowledge of how equipment should be operated and poor technical expertise in managing and fully utilizing equipment. (Figures 2-6 and 2-7.)

Learning to Use and Manage Equipment

The ability to *use* a piece of equipment fully is gained by studying manufacturing technology — learning the optimal conditions for equipment and auxiliaries, for example, condi-

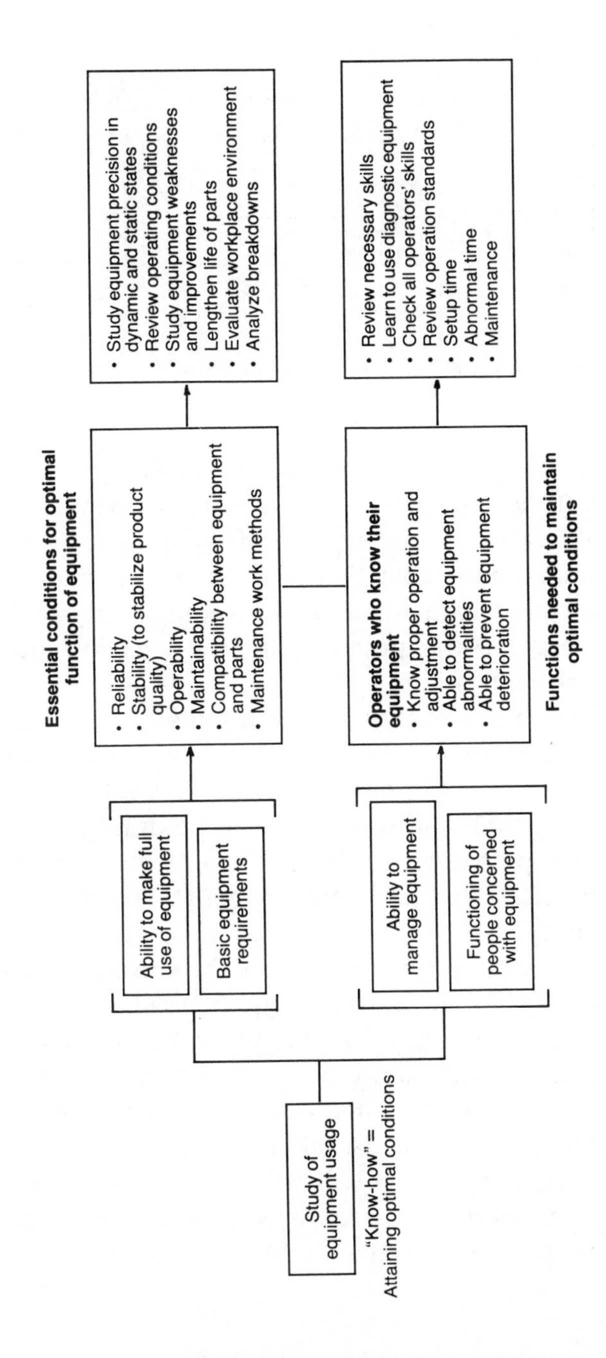

Figure 2-6. Study of Equipment Usage

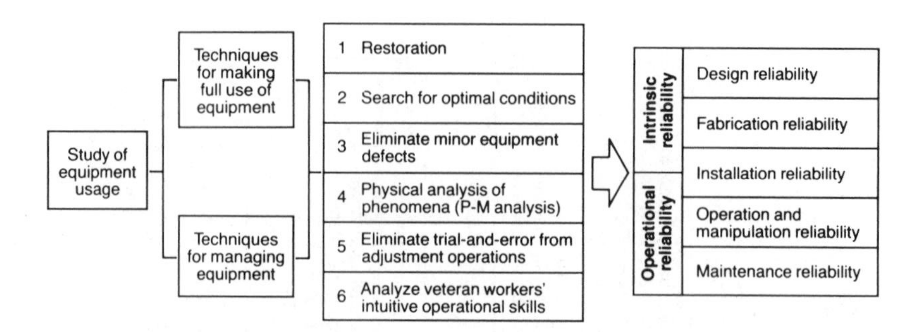

Figure 2-7. Learning to Use Equipment Fully and Make Basic Improvements

tions essential to enhance quality characteristics, to increase availability (time and speed), and to improve operability.

The ability to *manage* a piece of equipment is gained by studying human functions, that is, what people concerned with equipment can do to maintain optimal equipment conditions in terms of operation, manipulation, and the detection and restoration of abnormalities.

Even when the level of technical know-how is high, problems can surface when users ignore fundamental operational requirements or have poor operational skills. On the other hand, good operating skills and adherence to procedures is of little avail when the equipment itself is faulty. Clearly, the technologies of equipment utilization and management should be pursued simultaneously.

Equipment can be purchased, but the know-how necessary for its proper operation is not so easily obtained. It becomes available only when the company commits to training everyone involved in the use and maintenance of equipment. If training is inadequate, equipment is often neglected and operation is poor because of frequent problems. When new, sophisticated machinery is purchased, operators will be slow to master its operation since they lack fundamental knowledge and training. Problems similar to those experienced with the old equipment are likely to develop before long.

Restoration

All equipment changes slowly, over time. The timing and extent of the changes depend on the particular features of the equipment and its component parts. Often the changes are extremely small and easy to overlook. Large changes that cause unexpected breakdowns when they are not restored immediately are rarely neglected. When small changes are repeatedly neglected, however, they, too, can develop into major breakdowns over time. Even if they don't lead to breakdowns, they cause chronic losses. These changes are referred to as *deterioration*.

When equipment breakdowns recur in short cycles, the remedies sought often involve substitutions or replacements in mechanisms, parts shape, or material. These efforts are rarely effective because the breakdowns are not directly caused by the mechanism, part shape, or material, but by the neglect of smaller changes in conditions, such as abrasion, finishing precision, assembly methods, and precision. Only when these conditions are restored can breakdowns be prevented.

Restoration means returning the equipment to its original, proper, or ideal conditions (Figure 2-8). Since restoration prevents breakdowns by treating their fundamental causes, it should be performed before changing mechanisms or parts. If restoration does not eliminate the breakdowns, then efforts should be made to improve the equipment. This does not apply, however, to equipment that cannot satisfy current technical or market requirements.

Equipment can function to its full potential only when parts and component strength and precision are balanced. Restoration attempts to regain this balance throughout the equipment. If only some parts are restored and altered, losses will continue to occur.

Accelerated Deterioration

There are two types of deterioration: natural deterioration and accelerated deterioration. *Natural deterioration* is the normal wear-out that occurs in spite of proper use and maintenance.

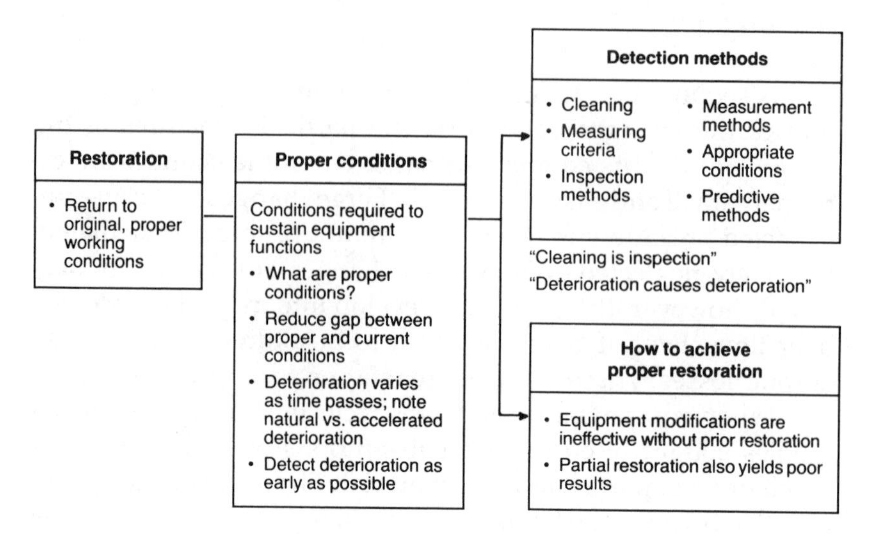

Figure 2-8. Restoration

Accelerated deterioration is caused by human factors and occurs over a much shorter period. It results from neglect of vital requirements for maintaining equipment functions, such as cleaning or essential lubrication. It also results from neglect of natural deterioration.

When neglected, deterioration tends to increase over time and spread to other parts. Indeed, unchecked deterioration can trigger a chain reaction that leads to an avalanche of problems. This situation is not uncommon on the shop floor. For example, a single loose bolt may cause some vibration. If it goes unchecked, however, the vibration gradually increases and other bolts begin to loosen.

Obviously, deteriorating conditions must be identified through inspection and corrected as soon as possible (Figure 2-9). Unfortunately, efforts to halt deterioration and restore equipment to its original operating conditions are often hindered because the following information is unavailable:

- original, optimal conditions
- methods for detecting deterioration
- criteria for measuring deterioration
- appropriate restoration procedures

Such problems can be avoided by establishing criteria and procedures in advance.

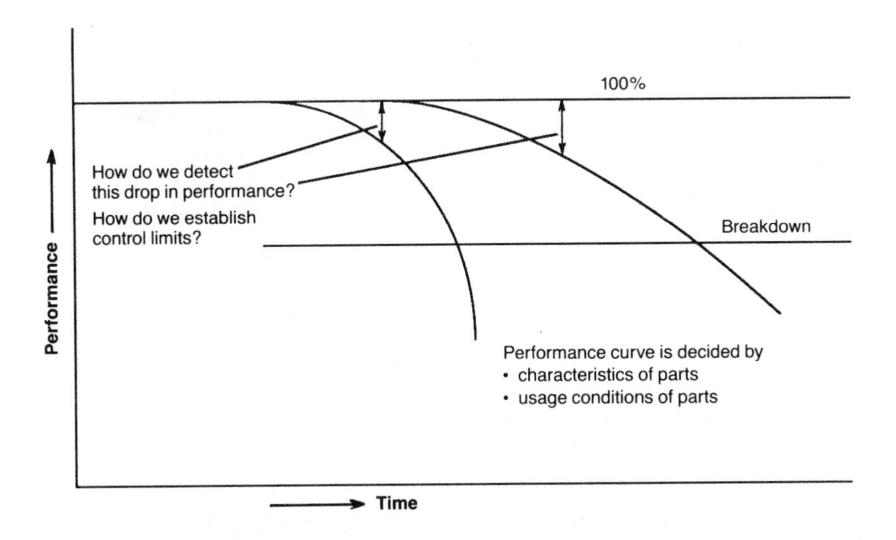

Figure 2-9. Development of Deterioration

Cleaning

Cleaning is an effective way to check and control equipment deterioration. JIPM encourages companies to use cleaning as a primary form of inspection for several reasons (Figure 2-10):

- During cleaning, each part of the equipment is touched or handled.
- In the course of touching each part, the worker can discover problems such as overheating, vibration, abnormal noises, looseness, and so on.
- Removing dust, dirt, and grease (and applying proper lubrication) slows deterioration.

Cleaning is inspection to detect deterioration; it lengthens component parts life and maintains equipment precision and quality requirements (Figure 2-11).

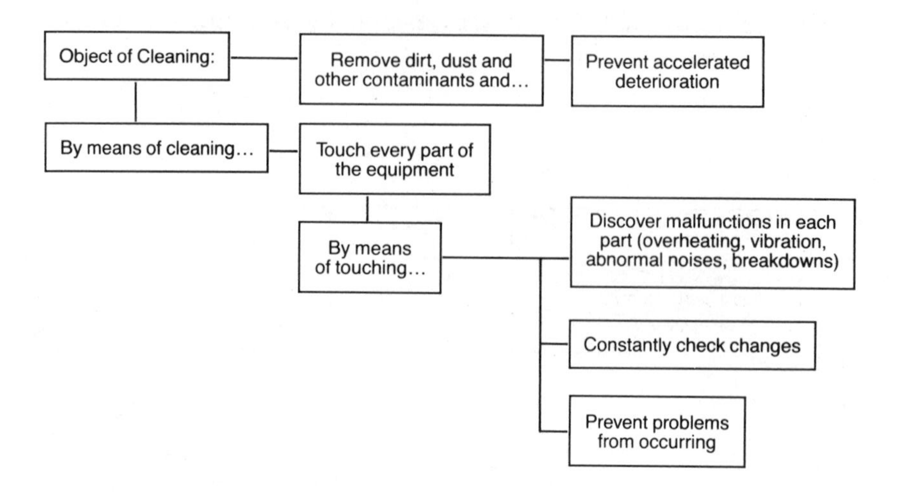

Figure 2-10. Cleaning Is Inspection

Physical Aspects	Psychological Aspects
1 Quality • Reduce quality defects • Stabilize quality **2 Equipment** • Detect malfunctions early • Prevent abrasion • Lengthen parts life • Maintain equipment functions • Prevent misoperation • Maintain parts precision	• Develop ability to detect malfunctions • Promote respect for equipment • Adhere to rules (discipline) • Increase motivation • Work in clean, sanitary workplace • Enhance purchaser's confidence

Figure 2-11. Effects of Cleaning

Consider, for example, the difference between washing a car in an automatic car wash and washing it by hand at home. In terms of gross appearance, the effects are the same: the car is clean, that is, dirt has been removed. Other less obvious physical defects, however, are revealed only by personal inspection of the various parts (*e.g.*, tire wear-out, presence of nails, cracks in the finish, body scratches, rust, etc.). This type of cleaning or inspection of factory equipment is important for the same reasons.

Predictive Maintenance

Deterioration can be detected through predictive maintenance as well. Diagnostic techniques measure the typical chemical and physical indications of the extent of deterioration in equipment. They also regularly compare current and normal operating conditions. If conditions are beyond the specified control limits, corrective measures are taken (*e.g.*, an overhaul or exchange of parts) to prevent a breakdown.

To conduct predictive maintenance, the following information must be known:

- how to measure deterioration
- how to detect signs of abnormality
- what "normal" conditions are
- where the line between abnormality and normality lies (when to intervene)

For example, to detect deterioration in high-speed bearings, vibration and other indications are periodically measured and compared to normal levels. If this comparison reveals a certain deviation from the normal level, an appropriate time for replacing the bearing can be estimated or predicted.

Optimal Conditions

Optimal conditions are those essential for optimal functioning and maintenance of equipment capabilities. Often, standards for parts and units are not established. Even when standards are available, there may be problems resulting from the way they were developed, or they may be simply ignored. When equipment is operated without understanding optimal conditions, the breakdowns and defects that occur are slow to be corrected.

By comparison, when individual units or parts of equipment are maintained at optimal levels (determined in accordance with appropriate engineering principles or functionally by observation), the equipment can be fully utilized over an extended period of time.

What Are Optimal Conditions?

Consider what is involved in determining the limits of precision and use for a single part in a machine: What are the dimensional control limits? What shape is acceptable? How much abrasion is allowed? All too often, tolerances of precision and control limits for operation are not established or are neglected, and the equipment is operated without understanding these requirements. Only by determining the conditions mentioned above and maintaining them based on function, mechanism, and usage of part can chronic problems and losses be eliminated (Figure 2-12).

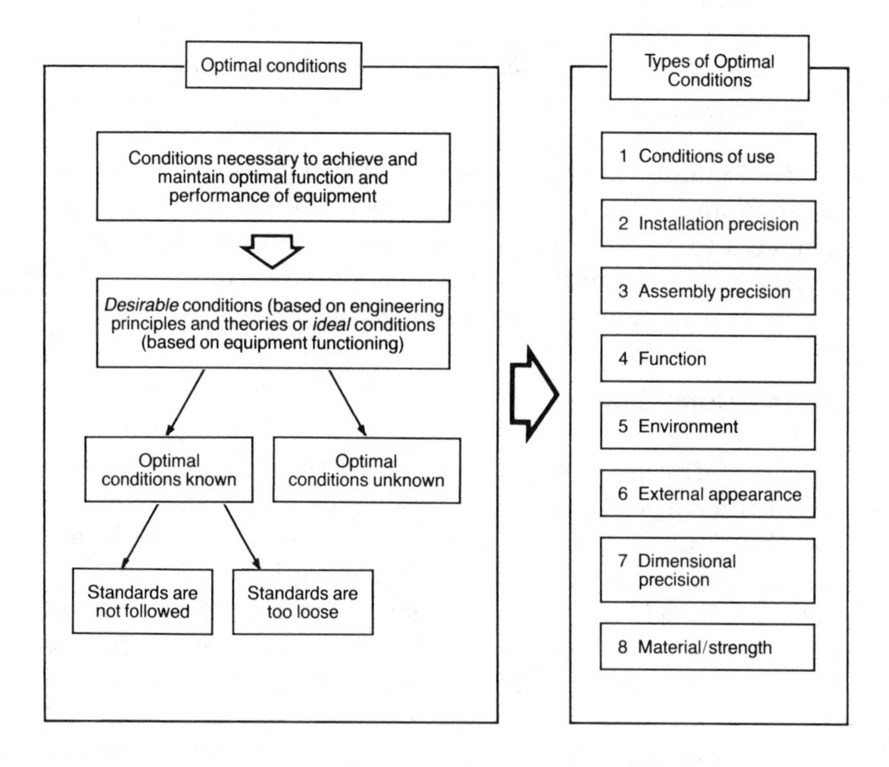

Figure 2-12. Optimal Conditions

Necessary Versus Desirable Conditions

The function of a V-belt, for example, is to transmit torque. What are the conditions for its operation? First, separate necessary from merely desirable conditions: *Necessary conditions* are the minimum requirements for operation; *desirable conditions* surpass the minimum or standard levels of operation.

Necessary conditions for the V-belt include

- conformance to specifications
- at least one belt installed (in the case of a triple-belt drive)

Desirable conditions include

- three belts installed
- all three belts at the appropriate tension
- no scratches, dirt, cracks, abrasion, grease marks
- proper alignment between motor and speed reducers

Desirable conditions are not directly linked to breakdowns and defects, but they influence the process indirectly. We tend to focus only on the necessary conditions, but desirable or optimal conditions should also be identified and maintained, since their neglect (*i.e.*, inadequate cleaning) often leads to the occurrence of chronic losses.

Establishing Optimal Conditions

Optimal conditions can be used to discover defective conditions. The gap created when current conditions are compared with optimal values highlights areas needing improvement. The following questions can be used to identify optimal conditions (Figure 2-13):

- *Dimensional precision:* Are parts accurately machined and measured?
- *Outer appearance:* What is the condition of the surface of parts and units (*e.g.*, dirt, seizure, uneven abrasion)?
- *Assembly precision:* If parts or components are precise, how is the precision of the integrated assembly? Poor integration can result from either poor assembly or faulty parts.

- *Installation precision:* Is equipment installed properly? Does it shake? Is it level?
- *Operational precision:* Are operating conditions optimal? Are the processing and manipulation conditions optimal? Be aware that operational conditions, in general, are simply based on past experience, without regard to what the optimal levels might be. Don't assume that "normal" is the same as "optimal."
- *Functioning parts:* Are critical parts functioning properly? Is actuation normal? Are the parts compatible with the equipment and system? Is the tolerance reasonable?
- *Environment:* Is the equipment environment favorable? Is the atmospheric temperature appropriate? Is the area free of dust and dirt? Are there other environmental requirements to maintain equipment?
- *Materials/Strength:* Is the material adequate? Is a more durable material available? Is the rigidity or strength sufficient?

When Optimal Conditions Are Not Known

Basic conditions such as those referred to above may not be known. Some information can be found in vendors' specifications, drawings, instruction manuals, and other technical resources, but it may or may not be adequate. Often, detailed parts, assembly instructions, or installation manuals are unavailable or misplaced. In such cases the equipment must be dismantled and analyzed in order to prepare in-house drawings and specifications. Examine and decide optimal conditions on a trial-and-error basis, and set control limits.

Differentiate Between Normality and Abnormality

Optimal equipment conditions are difficult to define when the boundary between normality and abnormality is ambiguous (Figure 2-14). Such cases are often neglected because

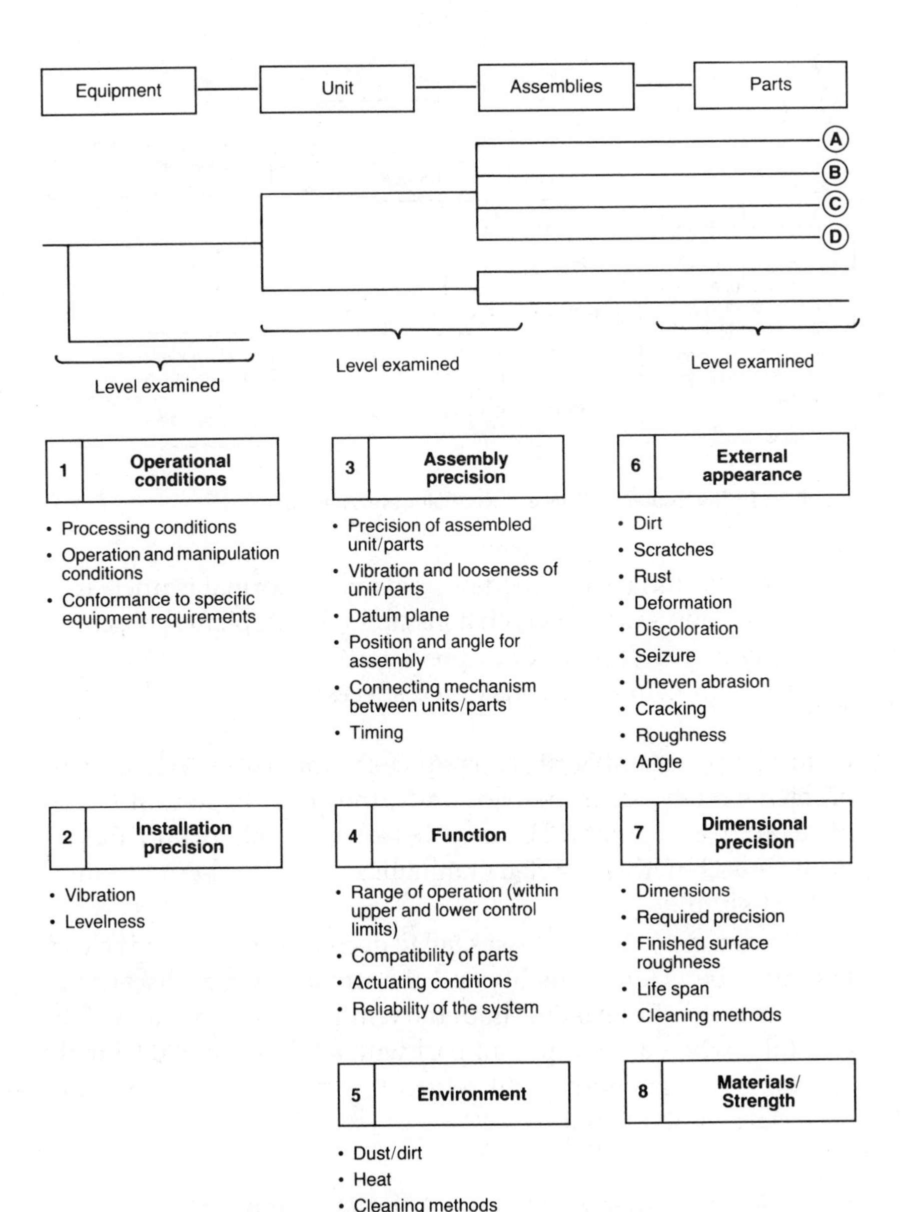

Figure 2-13. Establishing Optimal Conditions

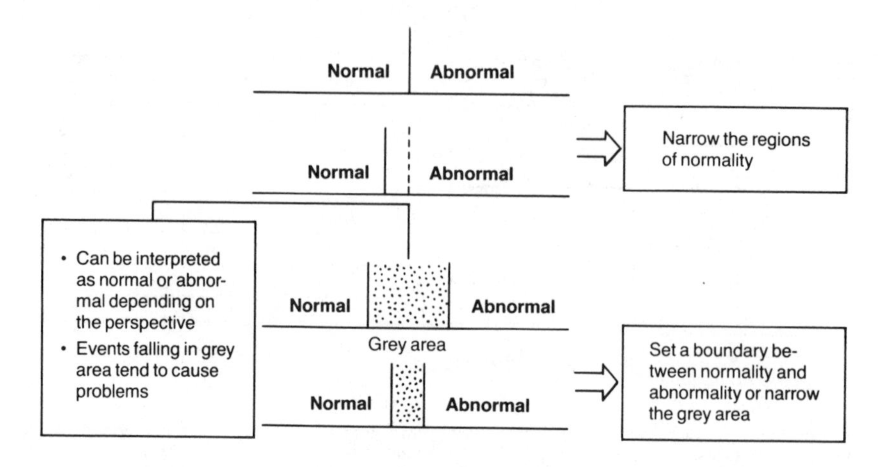

Figure 2-14. The Boundary Between Normality and Abnormality

- the boundary between normal and abnormal is unclear
- the cause-and-effect relationship in the problem is obscure
- repair results cannot be predicted
- results cannot be considered successful

Clearing up this ambiguity can reduce chronic losses. While such problems rarely occur in major units, they tend to go unnoticed in auxiliary equipment. The zone between normality and abnormality must be either clarified or minimized. Trial-and-error testing can be helpful.

Sometimes, chronic losses fail to decrease even when a clear boundary between normality and abnormality can be discerned. In such cases the existing boundary must be reevaluated. For example, when an equipment part with a tolerance of 0.05 malfunctions, production might return to normal when the tolerance is upgraded to 0.03.

Focus Targeted Area and Scope of Investigation

Before conducting detailed analyses to identify optimal conditions, think carefully about where and how far to look. It is difficult — and unnecessary — to check every single component part. Focus the investigation on only those areas directly related to the problem. Factors may vary according to

- the details of a specific problem's occurrence and its physical analysis
- the relationship between the occurrence and the equipment
- the machine mechanism and function/precision of component parts
- processing and operational conditions

These factors must be reviewed thoroughly and systematically, from both theoretical and technical perspectives. In TPM product quality is determined by machine quality; the statistical approach common in quality control studies should be avoided. These factors must be reviewed thoroughly from both theoretical and technical perspectives (Figures 2-15 and 2-16).

Changing Conditions to Expose Latent Defects

Changing conditions can help expose latent defects. For example, when equipment speed is raised above the current level, increases in the quality defect rate and decreases in the operating rate (*e.g.*, stoppages and breakdowns) prevent an actual increase in speed. The defects, breakdowns, and vibration that occur when the speed is raised are the result of poor precision of parts and components and their assembly conditions.

That such defects do not occur at the present speed and level of precision is irrelevant. What should interest us is their latency. When such defects eventually surface they not only display a cumulative effect but are often magnified exponentially (Figure 2-17). Therefore, raising the speed helps determine the optimal conditions for parts and components by prompting a reevaluation of their current precision.

Chronic Losses and Slight Equipment Defects

Equipment defects are not clearly defined, generally, but they are often divided into three broad categories:

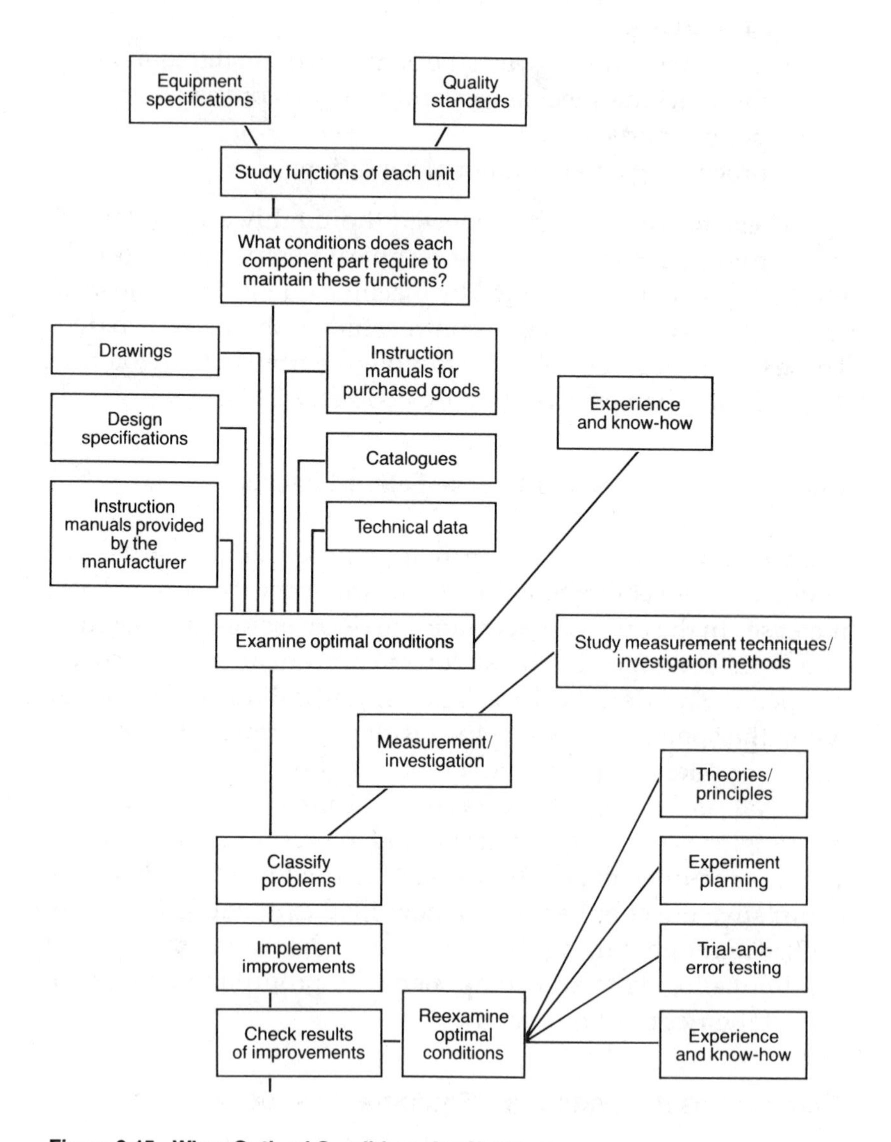

Figure 2-15. When Optimal Conditions Are Not Known

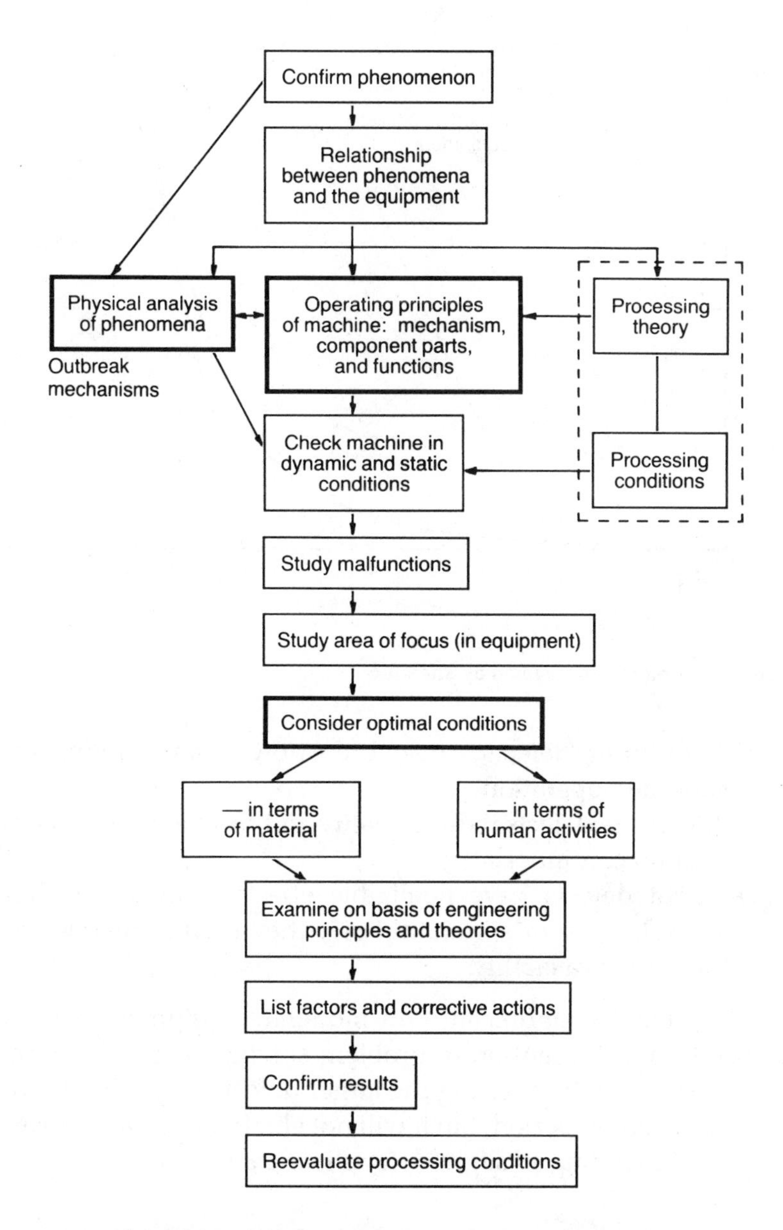

Figure 2-16. Relationship Between Phenomena and the Equipment's Optimal Conditions

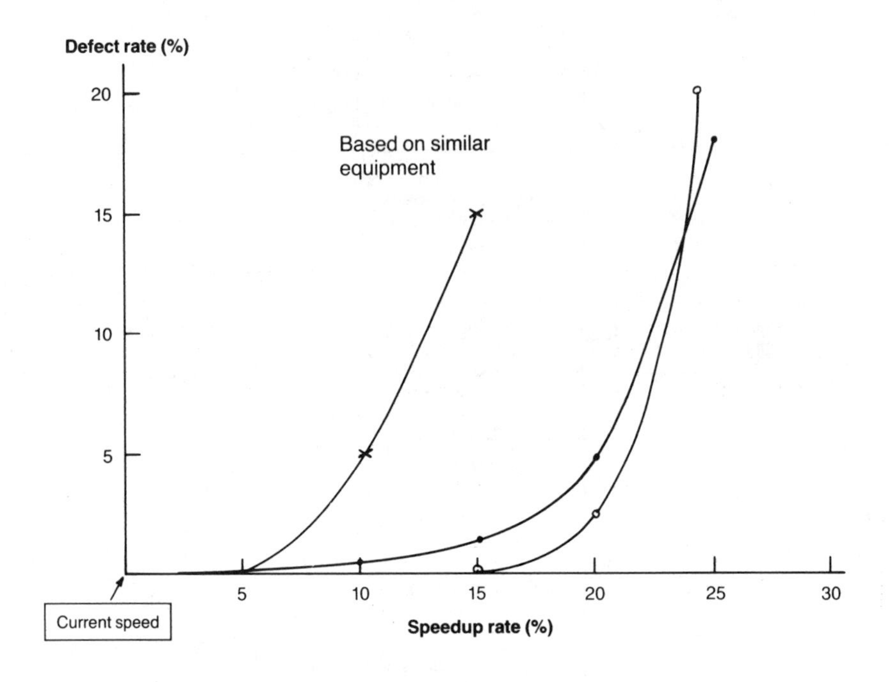

Figure 2-17. Defects Aggravated by Speedup

- Significant defects cause breakdowns and stoppages, stopping operation.
- Moderate defects reduce quality and productivity, but operation continues.
- Slight defects have negligible effects individually, but cumulatively or simultaneously they reduce quality and machine availability.

Traditionally, significant and moderate equipment defects receive the most attention in problem-solving, while slight defects are neglected. Targeting the larger defects may be effective in the early failure period, but it will not eliminate chronic losses. For this, eliminating *all* slight defects is the key.

Slight Equipment Defects

Slight defects are traditionally considered harmless, because individually their effect on breakdowns and quality defects is minimal. Dirt, grime, vibration, and 1 to 2 percent abrasion fall into this category, for example. Who imagines that a shift from 3 to 4 percent abrasion will cause an immediate breakdown? In general, however, slight defects include any suspicious factors that appear to have an effect on the result — regardless of their probability (*e.g.*, .01 to .001 percent).

Prevent dramatic cumulative effects. One important object of focusing on slight equipment defects is to prevent the potentially dramatic effect they produce cumulatively. Each defect should be dealt with thoroughly and patiently, because the overall effect is often greater than the sum of the individual defects. Even when the individual factors are extremely small,

- they can trigger other factors
- they can overlap with other factors to magnify the effect
- combined with other factors they can cause a chain reaction

Highlight causes. The second object in focusing on slight defects is to highlight causes by uncovering clues that may point to solutions for chronic problems. The cause-and-effect relationships of chronic problems often remain vague despite persistent analysis of recurring phenomena, experimental plans, or quantitative analysis. If slight defects are involved, a different method must be used to determine what causes are present and what to do about them.

Although all problems have both direct and indirect causes, human intellectual processes alone cannot always discover them. In other words, we can rarely "think" our way to a solution. The analytical approach fails whenever a single cause, multiple causes, or primary and secondary causes are overlooked or mistakenly eliminated as insignificant (Figure 2-18). A better approach is to assume that cause or causes are unknown, because

- a single original cause may trigger numerous other factors
- there may be multiple causes
- the combination of causes may change with every occurrence

Eliminate all slight defects. Since no purely analytical approach can guarantee that we have identified precisely which defects are causing various problems, the more practical solution is to eliminate *all* potential slight defects.

Obviously, when causes are distinct and independent, an approach that investigates and proves specific hypotheses can be more efficient. When there are numerous and minute causes, however, the slower method recommended here is more certain to eliminate chronic equipment defects. Although time-consuming, this approach has proven highly effective in eliminating chronic problems: even if the problem is not solved immediately, it reduces the complexity of the phenomenon (by eliminating contributing factors) and exposes hidden conditions that may lead to a solution.

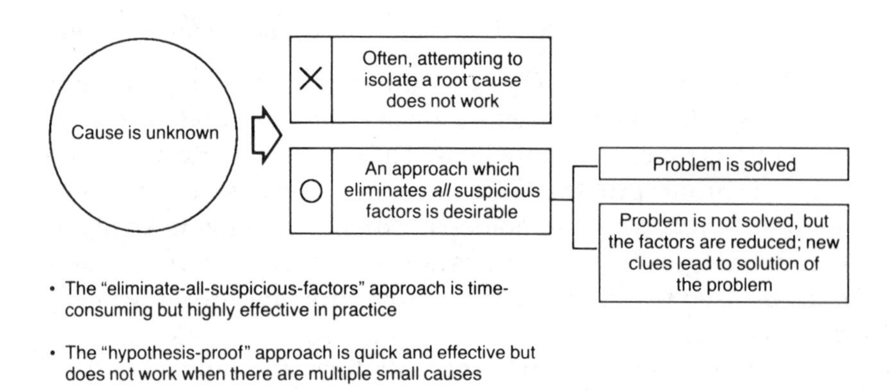

Figure 2-18. Approaches to Elimination of Chronic Losses

Prevent significant equipment defects from developing. The third aim of focusing on slight defects is to prevent small problems from turning into larger ones. If slight defects are ignored in their early stages, they often develop into large and fatal defects that cause breakdowns and major problems. Therefore, they must be removed as quickly as possible. Even when they do not lead to major problems, they contribute to accelerated deterioration and occurrence of moderate defects.

Eliminating Slight Defects

Keep the following two principles in mind when attempting to eliminate slight defects.

1. Evaluate the relationship between slight defects and equipment. When looking at slight defects from an engineering perspective, be sure to review the relationship between the defect and the equipment. This does not require a sophisticated theoretical analysis but a simple rethinking of basic principles and a review of all related factors.

2. Do not be misled. Keep in mind that the probability of any single defect's contribution to the overall problem is basically irrelevant. A defect that plays a minor role, for instance, may occur frequently. In any case, once the defect has occurred it ought to be eliminated. In other words, regardless of our preconceived notions about their importance, all suspicious defects must be eliminated, whether large or small. The size of the role is important only during the early failure period, when there is a high rate of defects and breakdowns. In such cases, it is most effective to identify and eliminate major defects first. The individual contributions of slight defects to chronic losses are almost impossible to determine, however, so targeting slight defects during this period will not be productive.

Case Study on Slight Equipment Defects

At factory O, incidents of pyrolysis (carbonization) occurred in a vinyl chloride extrusion process. Carbonization occurs when resin in the cylinder head is overheated. The equipment was designed to continuously regulate the flow of heat, but carbonization occurred when an abnormality in the flow of resin produced partial clogging.

Whenever this happened, the die, connected parts, and screws were disassembled and the carbonized part was cleaned so that operation could be restored to normal. Instead of improving, however, productivity continued to fall, and the operation staggered. To restore productivity, the following three measures were proposed to eliminate certain slight defects that had been observed to contribute to the carbonization:

- Thoroughly clean each part.
- Increase the precision of each part.
- Increase the precision of the parts assembly.

At first the company was hesitant, stating that such measures "could not lead to good results," "have little purpose," and "are a waste of time." The measures were implemented with reluctance. Gradual improvement encouraged operators to implement the improvement measures more diligently. Four months later, the problem had been entirely eliminated. Moreover, the causes were now understood:

- Thorough, regular cleaning of parts removed all dust or dirt that could quicken the conduction of heat and cause pyrolysis.
- Increased precision of individual parts eliminated scratches and abrasion that inhibited proper assembly and interrupted the flow of resin. Even if the individual parts had been normal, improper assembly had led to poor functioning and flow. This caused clogging and resulted in carbonization.

This case illustrates the tendency of companies to dismiss measures as ineffective because their potential effectiveness cannot be estimated. Such short-sighted thinking inhibits reduction of chronic losses. Implemented individually, the three measures might not have been effective. Together, however, they thoroughly prevented pyrolysis. Obviously, each measure was directly or indirectly related to the phenomenon of pyrolysis, yet each had only a minor effect on the overall result, and its relationship to the phenomenon could not be quantified. This example shows how important it is to control all variable factors simultaneously.

P-M Analysis and Chronic Loss

P-M analysis is a technique developed to promote the thorough, systematic elimination of defects that contribute to chronic losses. It provides a more effective approach than methods traditionally used in quality control activities.

Factor Analysis Inadequate for Eliminating Chronic Losses

Factor analysis (the cause-and-effect diagram) is a traditional quality control method used easily and effectively to solve many types of problems. Its weaknesses surface, however, when it is used to analyze the complex causes related to chronic problems. The major weakness in factor analysis is in the listing and treatment of factors.

Insufficient confirmation and classification. Conclusions are often based on insufficient observation and analysis. Although several events or phenomena may resemble each other, further observation often makes differences clear. It is important to consider carefully and compare the conditions under which the phenomena occurred and the results. If this process is incomplete, the wrong factors may be emphasized and ineffective remedial measures taken. Confirming the true nature of a problem is crucial.

Insufficient analysis of the phenomenon. When potential causes are listed without a practical, systematic approach, unrelated factors may be included and vital factors overlooked. An approach based on sound physical analysis is required.

Factor analysis often begins without a complete understanding of the physical conditions. The analysis is typically applied to a limited group of factors, while other factors are consciously ignored. It also does not consider information about similar cases, so the factors are often biased and limited. All efforts are focused on corrective action; the mystery of the phenomenon remains unsolved and the number of chronic losses does not decrease.

P-M Analysis Procedure

P-M analysis was developed to make up for these deficiencies in conventional analysis. *P-M* is an acronym of words starting with the letters *P* ("phenomena," "physical," "problem") and *M* ("mechanism," "machinery," "manpower," "material").

Through P-M analysis, all pertinent factors in a chronic loss are efficiently identified and eliminated. It includes the following steps:

Step 1: Clarify the Problem

Carefully investigate the problem and compare its appearance, conditions, and affected parts with those of similar equipment. Determine whether the phenomena are the same or slightly different.

Step 2: Conduct a Physical Analysis of the Problem

Consider the natural laws behind the phenomena observed. For example, when two objects come into contact, the weaker material will be scratched. Thus, if scratches occur frequently in a process, look for evidence of friction. Observing the points where the two parts come into contact clarifies the specific problem areas and factors.

This step is essential for several reasons:

- Physical analysis provides a unique perspective on the phenomenon and pinpoints causes.
- A logical and systematic investigation ensures that factors will not be overlooked.
- It discourages reliance on intuition and hunches.
- It forces us to reevaluate the basic causes, corrective actions, and control points for chronic losses unsuccessfully addressed in the past.

Step 3: List Every Condition Potentially Related to the Problem

Consider what conditions must be present to produce the phenomenon. Once these conditions are known, all situations that might develop into disorders can be prevented. A physical analysis of the outbreak mechanism allows us to study these underlying conditions systematically.

Under traditional factor analysis, corrective action addresses only some of the many contributing conditions. Losses cannot be completely reduced when some conditions are overlooked. Take care to avoid setting priorities or approaching the analysis with any preconceived ideas that may limit the analysis.

Step 4: Evaluate Equipment, Materials, and Methods

Consider each condition identified in step 3 in relation to such factors as the equipment, jigs and tools, material, and work methods involved. List all possible factors that influence the conditions. The relationship between these factors, the phenomena, and conditions must be made clear.

Step 5: Plan the Investigation

Carefully plan the scope and direction of investigation for each factor. Decide what to measure and how to measure it, and select the datum plane.

Step 6: Investigate Malfunctions

All items planned in step 5 must be thoroughly investigated. Keep in mind optimal conditions to be achieved and the influence of slight defects. Avoid the traditional factor analysis approach, and do not ignore malfunctions that might otherwise be considered harmless.

Step 7: Formulate Improvement Plans

On the basis of the preceding investigations, plan improvement strategies for each factor and implement.

Table 2-6 (see pages 76-77) is an actual example of PM analysis.

THE QUEST FOR SKILL

TPM developer and consultant Masakatsu Nakaigawa defines skill as the "ability to perform a task as if it were a reflex action." Skill assumes a level of expertise and experience that allows one to respond almost instinctively to problems; it is also an ability that is retained for a long time.

Manufacturing skill enables a worker to perform daily operations efficiently and correctly. It enables a person to judge a problem quickly and correctly, determine its causes, and take corrective action that will restore the status quo. For example, a skilled worker, hearing an unusual noise or vibration in the revolving part of machine, will decide whether it constitutes an abnormality, locate its source, and determine whether action is needed immediately or can be delayed. Skilled people react to such phenomena in the workplace reflexively.

Trained reactions become reflexes when they are used often. For example, a driver suddenly encountering an obstacle in the road automatically steps on the brake or swerves to avoid it. The speed with which the driver is able to respond determines whether an accident will occur or not. The more skilled the driver is, the shorter the reaction time will be.

Five senses	~	**Brain**	~	**Body**
↓		↓		↓
Perception	~	**Judgment**	~	**Action (a function of time)**

Raising Skill Levels through TPM

An important goal of TPM (and of factory improvement in general) is to raise workers' skill levels. The workplace will improve dramatically when everyone becomes more skilled. Skills of particular value in the workplace include

- attention (ability to observe and discern phenomena)
- judgment
- correct action and treatment
- restoration
- prevention
- prediction

Attention

The ability to concentrate and discover abnormalities requires attention and discernment. These skills are developed through basic education and training related to abnormal signs, judgment criteria, and detection methods.

Judgment

After discovering a problem, a worker needs the ability to think logically and make sound decisions. Making the wrong decision in response to an uncommon phenomenon is the result of inadequate knowledge and training. Many problems in the workplace are caused by undeveloped judgment skills.

Action, Treatment, and Restoration Skills

Prompt, appropriate, and informed action is valuable in any situation. The worker should always be able to restore the conditions within the time specified, with minimum losses. When restoration skills are undeveloped, the worker cannot take appropriate corrective action. This exacerbates the problem or creates a chain of errors.

Factory	Phenomenon	Description	Basic conditions	Relevance of equipment, materials, jigs, and tools
Extrusion process of vinyl chloride	Pyrolysis	Carbonization caused by excessive heat; carbonization accompanied by partial clogging caused by abnormal flow	1. Gap between cylinder and screw	• Cylinder abrasion
			2. Causes in the screw	• Eccentricity of the screw • Scratches on the screw • Screw abrasion • Dirt on the screw
			3. Assembly precision of individual parts	• Assembly precision
			4. Precision of individual parts	• Dirt in surrounding area

Table 2-6. Example of PM Analysis (1)

Factory	Phenomenon	Description	Basic conditions	Relevance of equipment, materials, jigs, and tools
Dry battery process	Batteries falling on revolving table	Loss of balance accompanying shift of center of gravity caused by external conditions (Shock, friction, shaking, etc.)	1. Conditions creating friction • Contact between revolving table and product • Factors inherent in product (warping of bottom, abnormal attachment)	Omitted
			2. Conditions creating shaking • Factors inherent in revolving table (undulating, involuntary shaking) • Contact between revolving table and surrounding guides	• Table surface conditions, levelness, and shaking • Irregular revolving • Guide shape, position, angle and surface conditions • Contact point conditions

Table 2-6. Example of PM Analysis (2)

Preventive Skills

A worker must be able to prevent as well as correct problems. When basic preventive knowledge and methods are lacking, preventable problems are often overlooked. For example, many problems can be prevented when the proper conditions for jigs and tools are known (*e.g.*, required precision, assembly methods) and when measurements are performed periodically to check for abnormalities and the need to replace parts.

Prediction Skills

A problem may be predicted on the basis of subtle signs, minor occurrences, or even a slight suspicion. The operator who knows equipment well and observes it skillfully can detect and eliminate many potential problems before they become serious.

Why Skills Must Be Taught

Are the following statements true? *You can't do operation X unless you're a veteran. It takes time to become an expert. Production is unstable when the number of new workers is greater than the number of seasoned workers.*

Certainly, some complicated operations and jobs require high skill levels. On the other hand, equipment improvements and adjustments that have been streamlined and simplified eliminate the need for some skills. In most companies, questions like the following are rarely answered with any certainty: What is the difference between new and experienced workers? Do they make different kinds of mistakes? How do skilled workers differ? How does production differ?

In general, some operations are performed incorrectly whether new or experienced people are doing the work. Skills that need to be taught are not analyzed well enough to be presented; education and training are thus inadequate. Often, knowledge has been communicated but results and performance are not checked.

Skill differences emerge for different reasons:

Don't know. Workers don't know the principles behind proper equipment operation, adjustments, and corrective action against abnormalities. In other words, there is a lack of knowledge and training.

Can't do. Although workers understand proper equipment operation, adjustments, and corrective action in theory, they cannot make the operation run smoothly in practice. Sometimes it works, sometimes it doesn't. These problems are caused by a lack of training and practice.

Won't do. Workers have the necessary ability, but they fail to maintain the set standards. Their neglect develops from overconfidence; they change procedures arbitrarily. Generally, this type of problem is caused by low morale and supervisors' neglect rather than lack of education or training.

Aims of Skill Analysis

Skill analysis has the following aims:

- Review current work methods and distinguish between methods that need improvement and those that require special skills.
- Clarify *essential* skills (and purge superficial skills).
- Provide the basis for instruction manuals that can promote skill and speed up education and training.
- Prevent repetition of the same mistakes.
- Teach individual skills to all workers and standardize the quality of production.
- Prevent the decline of skill levels with periodic checkups. In other words, make sure everyone can deal with all types of problems and check those skills regularly.

The Four Stages of Skill Development

A worker passes through four stages in the process of becoming skilled:

1. Don't Know (No Education)

At this stage, workers' knowledge of work methods, equipment, and production principles is minimal. This is the lowest level of skill.

2. Have Knowledge, But Can't Perform (Education, But No Training)

Workers possess theoretical knowledge of equipment and production principles but have not been trained to put it into practice.

3. Can Perform, But Not Well (Insufficient Training)

The worker's performance is poor and inconsistent because of lack of practice and incomplete training.

4. Can Perform With Confidence (Fully Trained)

The worker has knowledge and training and is practiced in the work. Performance on this level is error-free and consistent.

Importance of Skills

Workers should be able to progress from one level of skill to the next with reasonable speed. Ideally, everyone must attain a high level of skill and confidence in their performance. Maintaining these levels over time will lead to dramatic improvements in the workplace.

To raise the level of skill, the following steps are crucial:

Organize Necessary Skills and Education

Knowledge plays a key role in enhancing skill. An educational program must organize the basic knowledge that workers need. Basic knowledge controls action; lack of knowledge leads to incorrect actions and operations, faulty cause analysis, and ineffective corrective action. It generates outbreaks of new problems and exacerbates old problems. As the degree of automation increases, the area of relevant basic knowledge also expands. Table 2-7 lists some examples of basic knowledge that should be taught in an education and training program.

What knowledge is required of the average worker?

1 Key points of work
2 Criteria for judgment
3 Check results of work
4 Major equipment design data
5 Operating principles of the equipment
6 Managing equipment based on the operating principles (key points)
7 Main component parts function and precision to be maintained
8 Method for measuring precision
9 Parts replacement and checkup methods
10 Adjustment methods
11 Control methods and system
12 Detection and treatment of abnormalities

★ If information is disorganized and training is not thorough, workers' lack of knowledge will result in:

- Incorrect actions and work
- Incorrect assumption about causes and ineffective corrective action
- Aggravation of existing troubles

Table 2-7. Examples of Basic Knowledge

Provide Training and Practice through Constant Repetition

"Knowing" is quite different from "doing." Training is based on the fundamental knowledge acquired through education. Training and practice transform that knowledge into skill (Figure 2-19). How much practice is required depends on the characteristics of the operation, but in every case it is important to practice diligently and patiently. With persistent practice, performance becomes accurate and efficient.

Typically, education is actively provided (although often incompletely) but rarely checked in practice. To enhance training

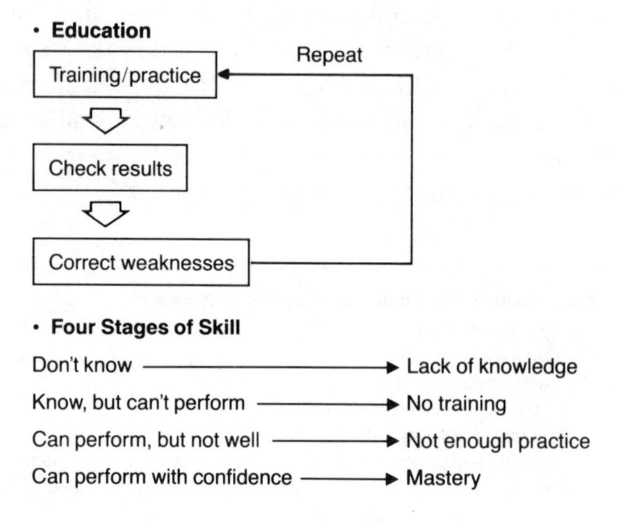

Figure 2-19. Training (1)

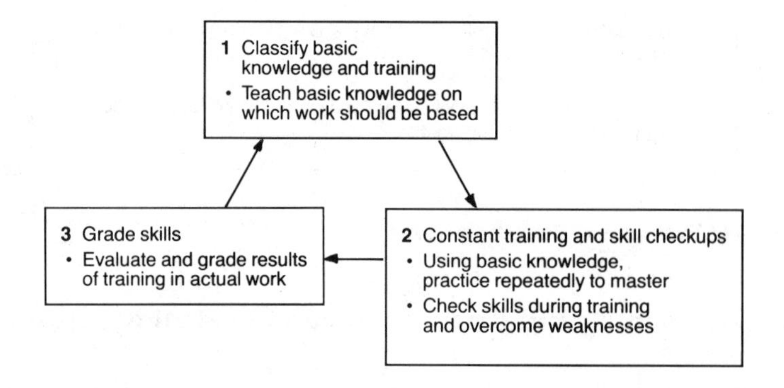

Figure 2-19. Training (2)

effectiveness, observe trainees to ensure that they are performing the required actions properly. If they are not, correct and check their work again (Figure 2-20). Moreover, checking the performance of operating procedures, sequences, standards, and other points may bring to the surface problems in the training

Constant Training and Skill Checkups

Training is based on fundamental knowledge acquired through education. Skills are mastered by repeating the action over and over again to increase speed and precision

Thorough and continuous training

Evaluation of Skills

- Check to see whether the operation is being carried out faithfully and accurately
- Don't check the results; check the process

1. Monitor training over a short period to hasten the development of workers' expertise
2. Observe the results of workers' self-training
3. Make operators aware of their own level of skill
4. Guide each worker in developing his/her own skills
5. Teach production line leaders how to guide their workers effectively

Figure 2-20. Building and Mastering Skills

program, machinery, and so on. If there are equipment problems, for example, the locations, causes, frequency, and conditions can be clarified and the information passed on to future trainees.

Evaluation of Skills

Check trainees' skill levels for accuracy and speed in production tasks such as setup and adjustment, treatment of abnormalities (quality and equipment), daily cleaning, inspections, lubrication, and so on. Then set standards for performance levels. For example, set level 1 at 90 percent or more (*i.e.*, for individuals who can perform 90 percent of the tasks), level 2 at 60 to 90 percent, and level 3 at less than 60 percent. Setting achievement levels gives individuals targets to work toward and improves the quality of instruction by requiring trainers to provide constructive evaluation.

REFERENCES

The ideas on equipment efficiency presented in this chapter are based on concepts developed by the Japan Management Association. The concepts of slight defects and measuring overall effectiveness were adapted from Masakatsu Nakaigawa, Japan Management Association.

Nakaigawa, Masakatsu. *Skill Management Text*. Tokyo: Japan Management Association, 1977.

———. "Skill management as a management system" (in Japanese). *IE Magazine* 20 (November 1978): 23.

———. "Skill management and changing attitudes in the factory" (in Japanese). *IE Magazine* 22 (March 1980): 28.

3

Eliminating the Six Big Losses

This chapter reviews TPM improvement activities aimed at eliminating the six big equipment losses: breakdowns, setup and adjustment time, idling and minor stoppages, reduced speed, process defects, and yield losses.

TAKING ACTION AGAINST BREAKDOWNS

Most people recognize that breakdowns are a major form of loss in manufacturing, but for a variety of reasons few companies do much to reduce the scope of that loss. To take this loss seriously and begin reducing it requires, first of all, new thinking about breakdowns.

Changing How We Look at Breakdowns

In Japanese the original meaning of the term *breakdown* (*kosho*) is "to deliberately destroy something old." In other words, *breakdown* meant damage caused by willful human actions. As this suggests, equipment breakdowns are often caused by human assumptions and actions.

Many people assume that

- it is not the operator's responsibility to perform inspection
- all equipment eventually breaks down
- all breakdowns can be fixed

It is not surprising, then, that breakdowns are difficult to eliminate. Eliminating breakdowns for all equipment is possible only if people change how they think about and use equipment.

Begin by Cultivating New Attitudes

First, people concerned with equipment must replace their assumption that "all equipment eventually breaks down" with the conviction that "equipment should *never* break down." Then everyone else, including operators, is more likely to accept the idea that equipment can be used in a way that actually prevents breakdowns. Moreover, when people accept the view that everyone is responsible for equipment, operators will want to learn how to use their own equipment so it won't break down.

Two Types of Breakdown

According to the *Japan Industrial Standards* (JIS), a failure or breakdown is the "loss of a specified function in a certain object (*e.g.*, system, machine, part)." Since the phrase "specified function" is vague, however, the meaning of *breakdown* remains elusive. Therefore, it helps to divide breakdowns into two categories: *function-loss breakdowns* and *function-reduction breakdowns*.

Function-loss breakdown. For most people, the term *breakdown* means a sudden, dramatic failure in which the equipment stops completely. Such unexpected breakdowns are clearly losses, because production is stopped. This is called a *function-loss* breakdown, or a breakdown in which all equipment functioning stops. Even if the cause lies in a single specific function, the breakdown results in the cessation of all equipment functions. Not all equipment failures are of this type, however.

Function-reduction breakdown. Deterioration (not failure) of equipment causes other losses even when the equipment can still operate. Long setup and adjustment times, frequent idling and minor stoppages, reduced manufacturing speed and cycle times, and increased defects in process and during startup are all possible losses of this type. Problems related to deterioration are considered *function-reduction* breakdowns, or breakdowns resulting in various losses (defects, minor stoppages, etc.). They are caused by deterioration in specific parts of the equipment and are considered less serious than function-loss breakdowns.

Generally, people tend to overlook function-reduction breakdowns. In many cases, however, function-reduction breakdowns account for the largest proportion of overall equipment losses.

Poor Equipment Management Promotes Chronic Breakdowns

Breakdowns become chronic for two reasons: organizational problems and technical problems related to the equipment. Figure 3-1, a relations diagram based on the experience of several companies, illustrates some common deficiencies in equipment management.

Organizational Weaknesses

There are many structural reasons for an ineffective response to the problem of chronic breakdowns. For example, in many production departments, operators accept the traditional strict division of labor between production and maintenance ("I operate — you fix") and have no interest in maintenance. In maintenance departments, workers are not adequately trained in the specialized skills needed to keep increasingly sophisticated equipment in good repair. Moreover, in engineering departments, overdependence on subcontractors and lack of time and money results in poorly designed equipment.

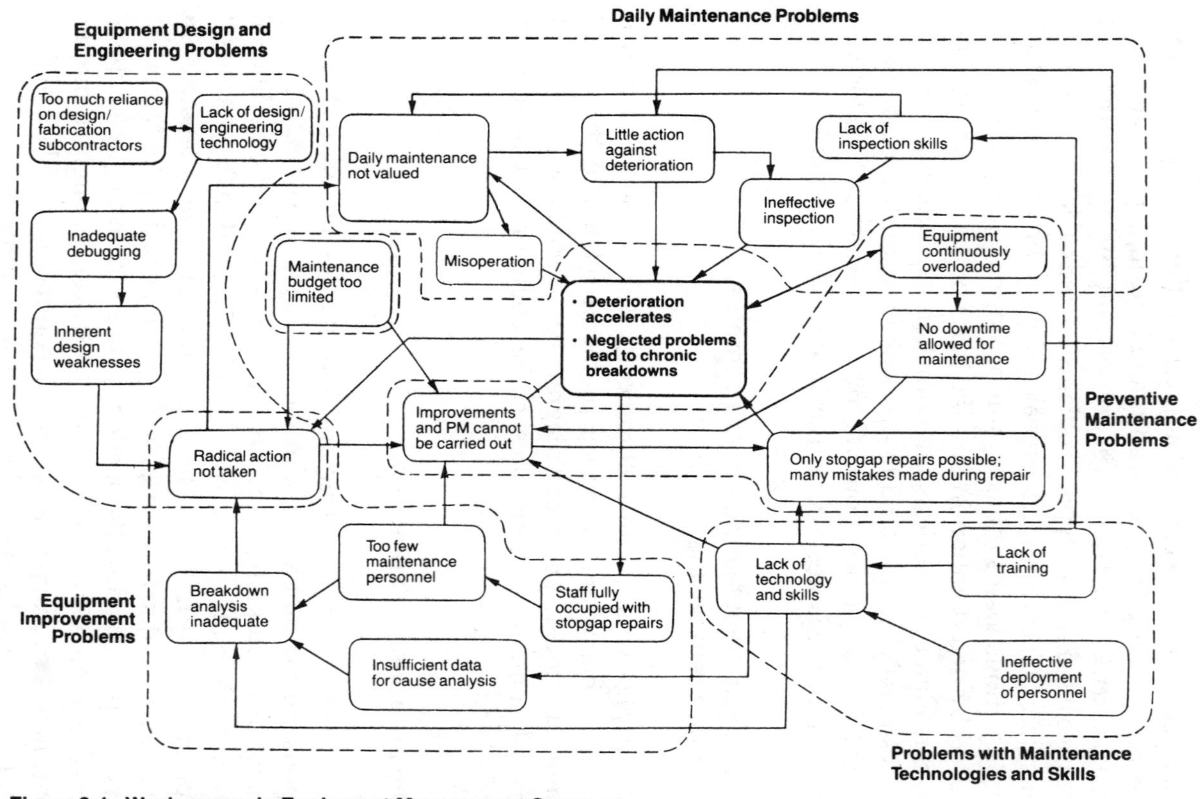

Figure 3-1. Weaknesses in Equipment Management System Result in Serious Problems

These problems occur when management is not sufficiently aware of the importance of productive maintenance. As weaknesses in the equipment and management structure become more and more evident, they produce a decline in morale and an increase in chronic breakdowns.

Hidden Magnitude of Equipment Losses

Management is often unaware of the scope of losses resulting from poor equipment maintenance. In addition to breakdowns, these losses include the five other major equipment losses: increased setup and adjustment time, frequent idling and minor stoppages, reduced speed and cycle time, deterioration in quality and yield, and energy and manpower losses. Often these losses make up 30 to 50 percent of processing costs.

Before improvement can begin, such weaknesses in a company's equipment management must be clearly identified. Breakdown reduction activities cannot be started until managers and supervisors understand the fundamental role of preventive maintenance in corporate improvement. Unless their support is assured, technical efforts will be useless.

Basic Principles for Zero Defects: Exposing Hidden Defects

Equipment defects are equipment disorders that cause breakdowns. Hidden defects remain invisible for one reason or another and untreated. Many breakdowns are caused by these unnoticed equipment defects; exposing and correcting them can reduce breakdowns to zero.

Breakdowns caused by hidden defects are chronic in nature, occurring so often that they seem "normal." For instance, a speed reduction loss in equipment is often difficult to discern, but it is a kind of breakdown or failure all the same — a form of chronic loss.

Hidden Defects Are a Major Cause of Breakdowns

Conventional maintenance efforts focus on sporadic, unexpected breakdowns and significant, highly visible equipment defects. Total productive maintenance, however, addresses both function-loss breakdowns and the function-reduction breakdowns caused by hidden defects. While a single, significant defect can trigger a breakdown, a combination of small hidden defects, which may seem completely unrelated to the breakdown (*e.g.*, dust, abrasion, vibration, loose bolts, scratches, warping), is often the major cause.

Such slight defects can develop into larger defects. Sometimes they overlap, magnifying the effect and triggering a dramatic loss (such as a sudden breakdown). Just as one unextinguished cigarette may cause a large fire, hidden defects spark breakdowns and should be stamped out while they are small. This is the fundamental principle behind preventive maintenance.

Defects Can Be Physically or Psychologically Hidden

Becoming aware of hidden defects is the first and most difficult step in eliminating breakdowns. Hidden defects do not have to be small to be hard to see. Even significant defects can be physically or psychologically obscured.

Defects can be physically hidden by

- poor inspection and analysis of deterioration
- poor layout and assemblies that are difficult to inspect
- dust and contamination

Defects can be psychologically hidden because

- defects are consciously ignored, even when visible
- the problem is underestimated
- the problem is overlooked, even though concrete symptoms are visible

For example, defects are hidden when equipment performance is poor but no effort is made to improve it. Suppose the equipment effectiveness rating reveals speed losses in the equipment, but both operations and maintenance say, "There's nothing we can do about this — it's the nature of the machine and the process." This defect will remain both physically and psychologically hidden until those involved acknowledge the opportunity and need for improvement.

Focus Attention on Hidden Defects

Eliminating breakdowns caused by hidden defects demands a new approach. If problem-solving efforts focus narrowly on the breakdown occurrence or on obvious individual defects, the wrong factors may be targeted, carrying the investigation even further from a solution.

Instead, focus improvement efforts on hidden defects and eliminate them as a class, since individually they are hidden. Because their contribution to losses is complex and continuous, it is not enough to simply treat or "fix" the visible effects of hidden defects. Thus, the purpose of autonomous maintenance and maintenance-prevention design is to foster an environment for equipment in which hidden defects simply cannot develop.

Stop Equipment for Inspection and Prompt Treatment

To eliminate defects — to expose and correct hidden defects — equipment must be stopped at reasonable intervals for inspection and maintenance. Production departments typically grumble (under the pressure of production) when the maintenance department requests a stop. The production lost from an hour's stop for inspection and servicing is minimal, however, compared to that lost in the dozens or more hours needed to treat a breakdown. In other words, the loss of production from planned stops can be turned into profit.

Five Requirements for Zero Breakdowns

Five types of action are necessary to uncover hidden defects and treat them properly:

- Maintain basic equipment conditions (cleaning, lubricating, bolt tightening).
- Adhere to operating conditions.
- Restore deterioration.
- Correct design weaknesses.
- Improve operating and maintenance skills.

1. Maintain Basic Equipment Conditions

Three factors are involved in maintaining basic equipment conditions: cleanliness, proper lubrication, and bolting (*i.e.*, keeping bolts and nuts properly tightened). Maintaining these basic conditions prevents equipment deterioration and helps eliminate potential causes of breakdowns.

Cleaning. Cleaning removes from equipment the dust and contamination that cause friction, clogging, leaking, defective running, electrical defects, and reduced precision in the moving parts. Thorough cleaning prevents the breakdowns, quality problems, and accelerated deterioration these defective conditions can produce.

Cleanliness requires more than superficial cleaning. Every nook and cranny of the equipment, jigs, and tools must be explored. This not only removes dirt and dust, but also uncovers hidden defects such as abrasion, loose nuts and bolts, scratches, overheating, vibration, abnormal sounds, and so on. In effect, *cleaning is inspection*. A trained worker can often find 200 to 500 hidden defects in the course of thoroughly cleaning a long-neglected piece of equipment. Significant defects, such as those listed in Table 3-1, may also be found.

Lubrication. Equipment cannot operate effectively without proper oiling and lubrication. In many factories, however, the reservoir or lubricator is left empty and is covered with dust and sludge. Often even the oil supply pipe is clogged or leaking.

Neglect of lubrication causes various losses (Figure 3-2). For example, it may be the direct cause of sporadic, unexpected breakdowns such as seizure. It also hastens equipment deterioration by causing abrasion and overheating, which affect the overall condition of the equipment.

	No.6 Intermixer	No.1 Banbury Mixer	No.6 D Extruder
Location	Attachment bolt on latch cylinder	Pedestal of outlet door	Pusher casing
Fault	Bolt snapped	Frame cracked	Casing cracked
Estimated losses due to delayed discovery	Replaced screw liner Shutdown time: 3 days Potential loss: $25,750	Fabricated and installed new pedestal Shutdown time: 4 months Potential loss: $70,000	Fabricated and installed new casing Shutdown time: 1 month Potential loss: $17,000
Corrective action taken	Bolt replaced and tightened	Repaired by metal-locking method ($3,000)	Temporarily repaired; new casing fabricated and installed ($3,750)

Table 3-1. Hidden Defects Discovered during Equipment Cleaning at Tokai Rubber Industries Plant

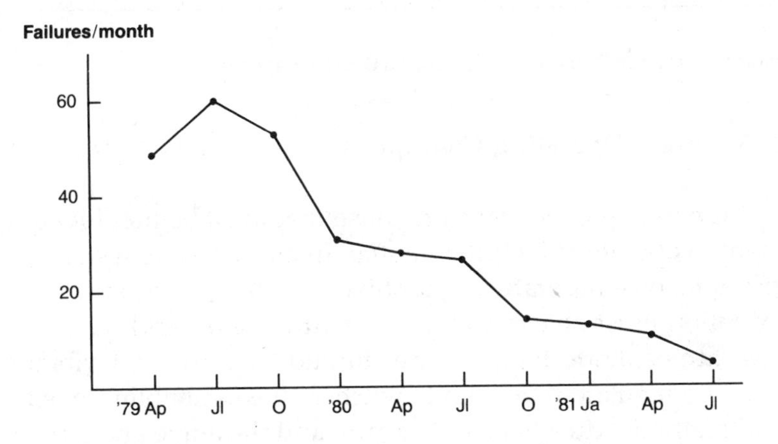

Figure 3-2. Reduction in Equipment Failures Due to Poor Lubrication (Topy Industries)

Bolting. Broken and loose connecting parts such as bolts and nuts play a large role in equipment breakdowns. A single loose bolt (*e.g.*, an assembly bolt in a bearing unit, die, jig, or limit switch or a flange bolt in a pipe joint) can directly cause a breakdown. Moreover, a loose bolt can increase shaking, which in turn loosens other bolts. This creates more vibration, triggering a chain reaction that can result in a serious breakdown before anyone is aware of a problem.

Faulty bolting is a common form of hidden defect. One company discovered that defects in bolts and nuts accounted for 60 percent of all breakdowns. Table 3-2 lists the hidden defects uncovered in an overall inspection.

Type of plant	Equipment	Total bolts checked	Fault			Total faults	Defect rate (%)
			Slack	Fallen out	Not engaged, etc.		
Belt	Vulcanizing presses (9)	10,494	2,651	89	267	3,007	28.6
Material	Weighing cutters (3 line)	2,273	1,053	38	—	1,091	48.0

Table 3-2. Hidden Defects Found through Overall Inspection

2. Maintain Operating Conditions

Operating conditions are those that must be met for equipment to operate at its full potential. In oil hydraulic systems, for example, oil temperature, quantity, pressure, purity, and level of oxidation must be controlled. In control panels and instrumentals, atmospheric temperature, humidity, dust, and vibration must be regulated. For limit switches, the assembling position and methods, the shape of the cam, and the angle and strength of the connection between the roller lever and cam must also be considered (Figures 3-3 and 3-4).

1. Reduction in Hydraulic Machinery Breakdowns

Month/Year	Oct/79	Oct/80
Hydraulic equipment breakdowns / Total breakdowns	58/434	8/88

2. Changes in Treated Oil and New Oil Used

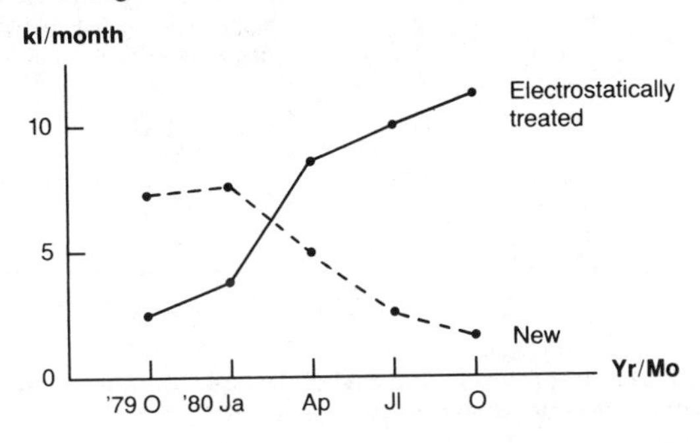

Figure 3-3. Effect of Electrostatic Cleaning of Hydraulic Oil

Unclear or incomplete conditions result in hidden defects. For example, if equipment is operating at below its standard speed, one or more hidden, nonstandard conditions may be the cause. To eliminate such defects, standard operating, manipulating, and loading conditions should be set and maintained for each piece of equipment and its individual parts. If equipment problems are treated without regard to these operating standards, the precision of operation and manufacturing conditions will not be stable, and breakdowns will continue to occur.

3. Restore Deterioration

Typically, when equipment breaks down, only the parts directly involved in the breakdown are restored; deterioration of the equipment, jigs, and tools is not treated. Therefore, even though a broken or worn part is replaced, breakdowns will recur

(Measured by replacements and adjustments per month)
Equipment monitored: heat-treatment lines Nos. 1-4
(Approx. 250 limit switches)

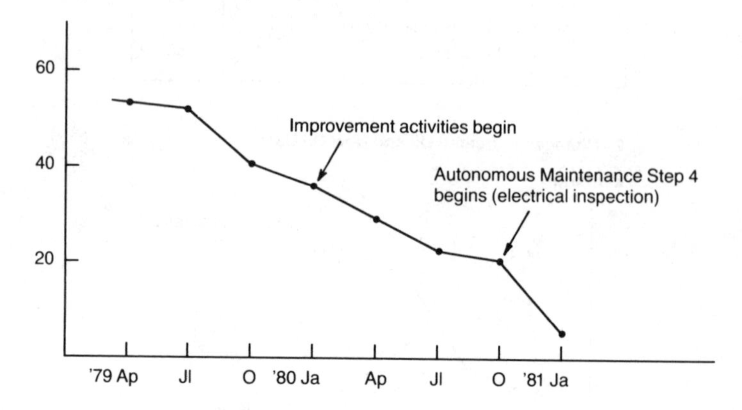

Figure 3-4A. Effect of Improving Limit Switch Usage Conditions on Their Breakdown and Adjustment

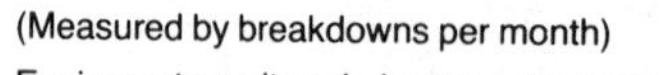

Equipment monitored: heat-treatment lines Nos. 1-4

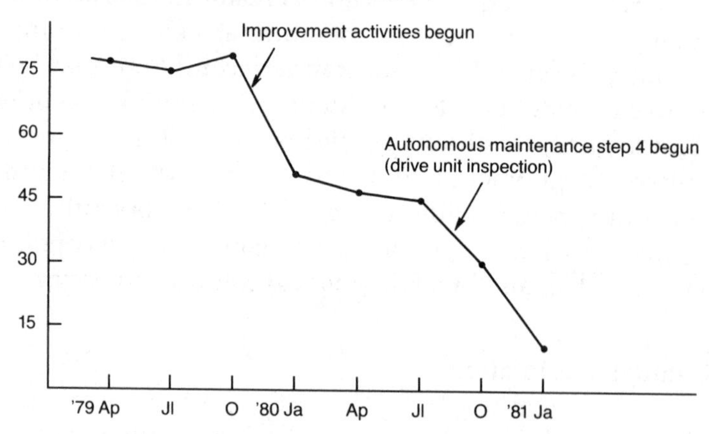

Figure 3-4B. Effect of Improving Usage Conditions of Sprockets and Chains on Their Breakdown Rate

because the balance of precision and strength in the equipment and tools has not been restored. Obviously, an imbalance caused by design or fabrication errors in machinery must be corrected by correcting design defects. If, however, the breakdown is caused in part by hidden deterioration of structural parts, partial restoration and design changes will not eliminate the breakdowns.

For example, if a driving shaft breaks off at a notch position in a machine, make sure that defects such as vibration caused by bearing abrasion, loose fit of the bearing, and backlash caused by abrasion between gears are considered and treated before changing the shaft or redesigning the notch.

Equipment slowly deteriorates over time and breakdowns occur as fatigues develop. Therefore, even if a broken part is restored and improved, breakdowns will continue to occur in other fatigued parts. At that point, before thinking about changes in design, return to the original drawings and use inspection and checkups to uncover deterioration.

Restoring the balance of equipment precision and strength on the occasion of equipment failure is an important strategy, but it is only a shortcut to breakdown elimination. To properly restore equipment, standardize the steps taken to discover and predict deterioration as well as the methods used to restore the deterioration. Discovery and prediction can be performed through periodic checkups, applying suitable inspection standards, and machine diagnostic techniques. Restoration carried out must be based on maintenance standards; it can also be realized through the skills and accumulated experience of maintenance personnel. (This is consistent with the TPM approach to preventive maintenance discussed in Chapter 5.)

4. Correct Design Weaknesses

Even when basic conditions (cleaning, lubrication, bolting) are strictly controlled, the costs of maintenance become enormous when equipment life is short and inspections, checkups, and restorative treatment cannot keep pace with the occurrence of defects. In such cases the problem may be caused by a design weakness requiring changes in equipment design, such as alterations in material, dimensions, and shape of component parts.

Avoid easy improvements. It is tempting to jump to conclusions, to oversimplify the problem (by assuming it is the same as other problems), or to rely strictly on instruction manuals without studying the breakdown data and equipment structure. This may lead to misinterpretation of the original design weakness and thus to failure of the alternative design. To comprehend the true source of the weakness and develop an improvement plan, do the following:

1. Take steps to properly understand the defective occurrence and the surrounding conditions (before and after the breakdown).
2. Confirm the equipment structure and functions.
3. Confirm the proper maintenance of basic equipment conditions, operation, manipulation, and loading conditions; confirm the restoration of deterioration on related functions.
4. Clarify the outbreak mechanism of the phenomenon.
5. Search for causes (design or other weakness, or both?).
6. Plan an improvement strategy.
7. Implement the improvement strategy.
8. Follow up and evaluate the results of improvement.

Correcting design weaknesses improves maintainability and lengthens equipment life (*see* Chapter 6).

5. Improve Operating and Maintenance Skills

In thinking about solutions to breakdowns, we tend to emphasize objects — equipment, jigs and tools, materials, and so on — and to forget the human factors. In fact, extensive education and training of operators, maintenance workers, equipment designers, and managers support any effort to achieve zero breakdowns.

Many breakdowns are caused by lack of skill. Human errors often go undetected, which makes them difficult to eliminate. The responsibilities of operators and maintenance workers must be clarified and their skill levels raised through education and training. Furthermore, TPM requires new ways of thinking about breakdowns and defects. Therefore, all education and training has two aspects: to improve skills and to improve understanding. Training to improve operating and maintenance skills is addressed in detail in Chapters 4 and 7 respectively; improving the understanding of everyone in the factory through small group activity is discussed in Chapter 8.

Pursue All Five Zero Breakdown Activities

All the five activities described above must be conscientiously pursued. Neglect of any one of them can directly trigger a breakdown; neglect of more than one area often causes malfunction in equipment indirectly and in hidden ways (Figure 3-5). Thus, one or two of these activities are usually insufficient against breakdowns caused by hidden defects. Even when several improvement strategies are used, breakdowns often continue to occur. Figure 3-6 (*see* pp. 102-103) reviews the five activities in detail. Remember that eliminating *all* hidden defects is the only way to eliminate *all* breakdowns.

Production Versus Maintenance

Why do breakdowns occur? At their root, breakdowns are the result of human factors — the erroneous assumptions and beliefs of engineers, maintenance personnel, and equipment operators. Breakdowns cannot be eliminated until those assumptions and beliefs are changed, particularly those regarding the traditional division of labor between production and maintenance departments. Operators and maintenance personnel must reach a mutual understanding and share responsibility for equipment. In fact, *everyone* concerned with the equipment must cooperate with and understand the role of everyone else. Each department must implement the actions against breakdowns mentioned in the previous section.

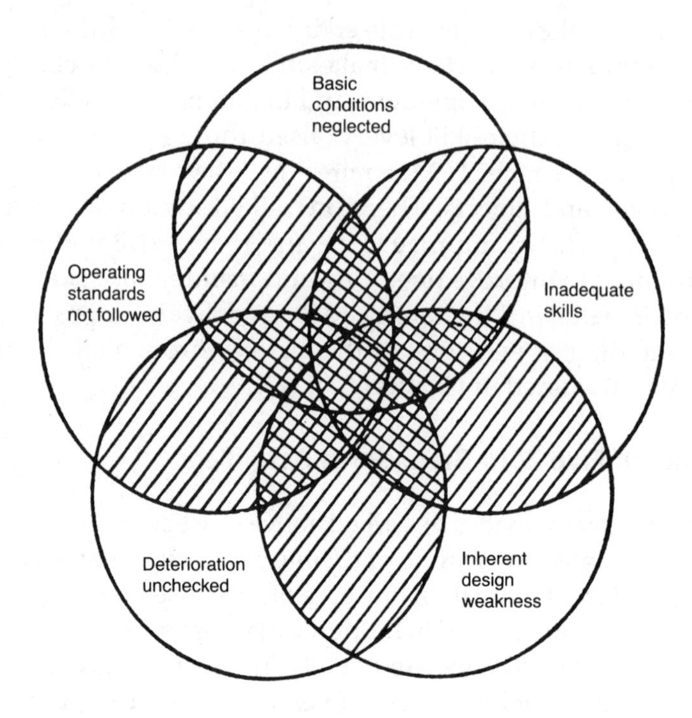

Figure 3-5. Combination of Breakdown Causes

Operators must do the following:

- Maintain basic equipment conditions (cleaning, lubrication, bolting)
- Maintain operating conditions (proper operation and visual inspection)
- Discover deterioration, mainly through visual inspection and early identification of signs of abnormalities during operation
- Enhance skills such as equipment operation, setup, and adjustment, as well as visual inspection

These activities constitute operators' autonomous maintenance responsibilities and are discussed in greater detail in Chapter 4.

Maintenance personnel must do the following:

- Provide technical support for the production department's autonomous maintenance activities
- Restore deterioration thoroughly and accurately, using inspections, condition monitoring, and overhaul
- Clarify operating standards by tracing design weaknesses and making appropriate improvements
- Enhance maintenance skills for checkups, condition monitoring, inspections, and overhaul

These activities have traditionally been responsibilities of the maintenance department (Figure 3-7).

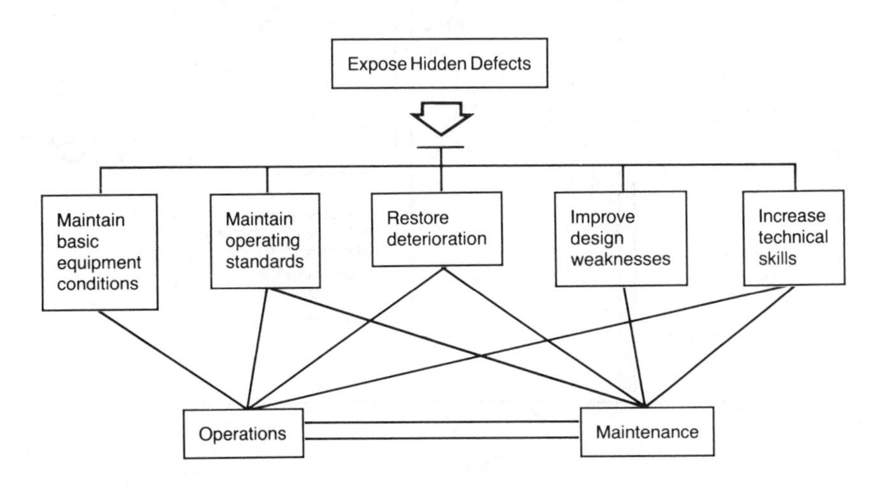

Figure 3-7. Role of Operations and Maintenance Departments

Zero Breakdowns: A Four-Phase Implementation Program

The five activities for zero breakdowns are not short-term programs, nor should they be implemented simultaneously. They are more effectively introduced in four consecutive phases (Table 3-3). Each phase has its own theme:

1. Stabilize equipment failure intervals (mean time between failures — MTBF).

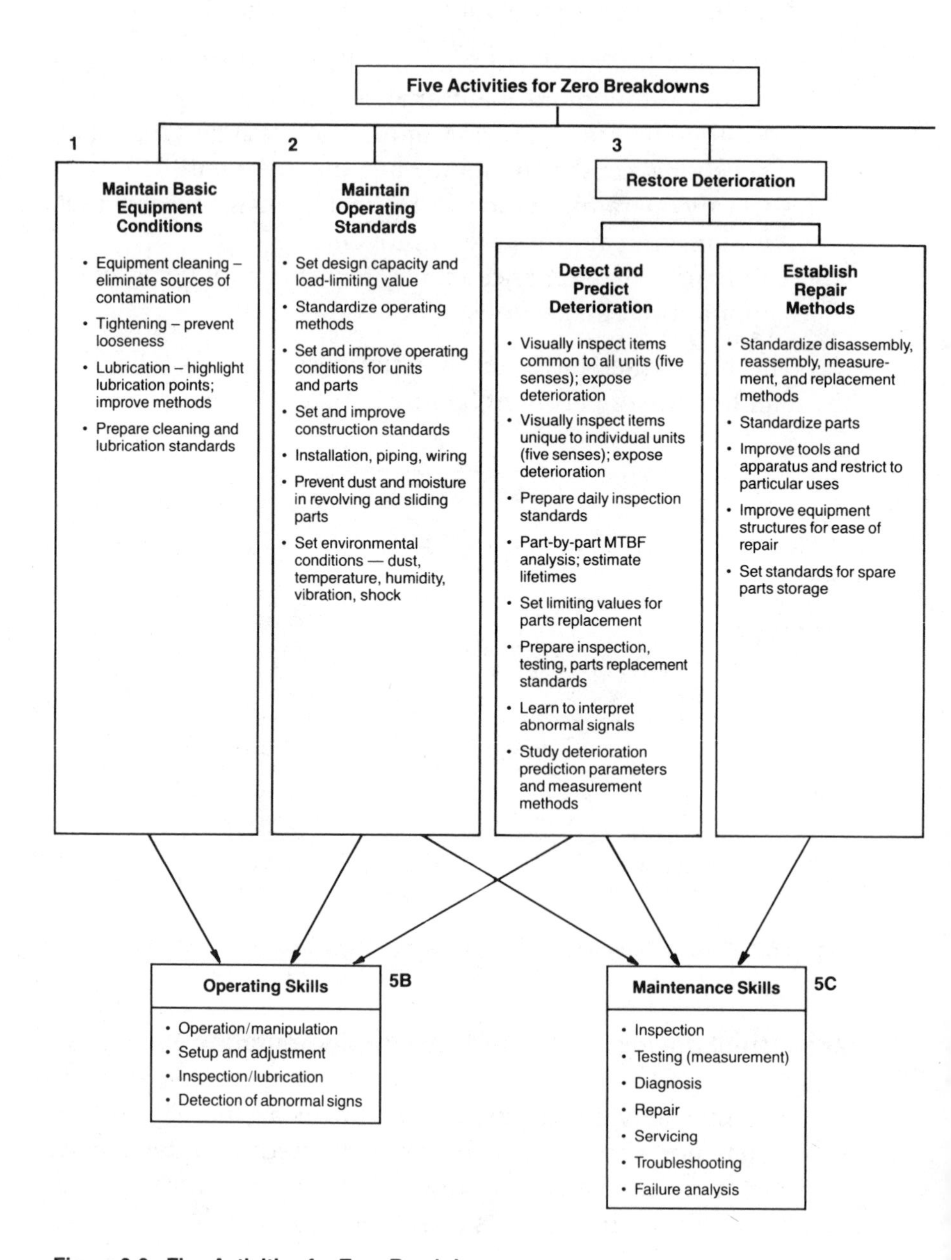

Figure 3-6. Five Activities for Zero Breakdowns

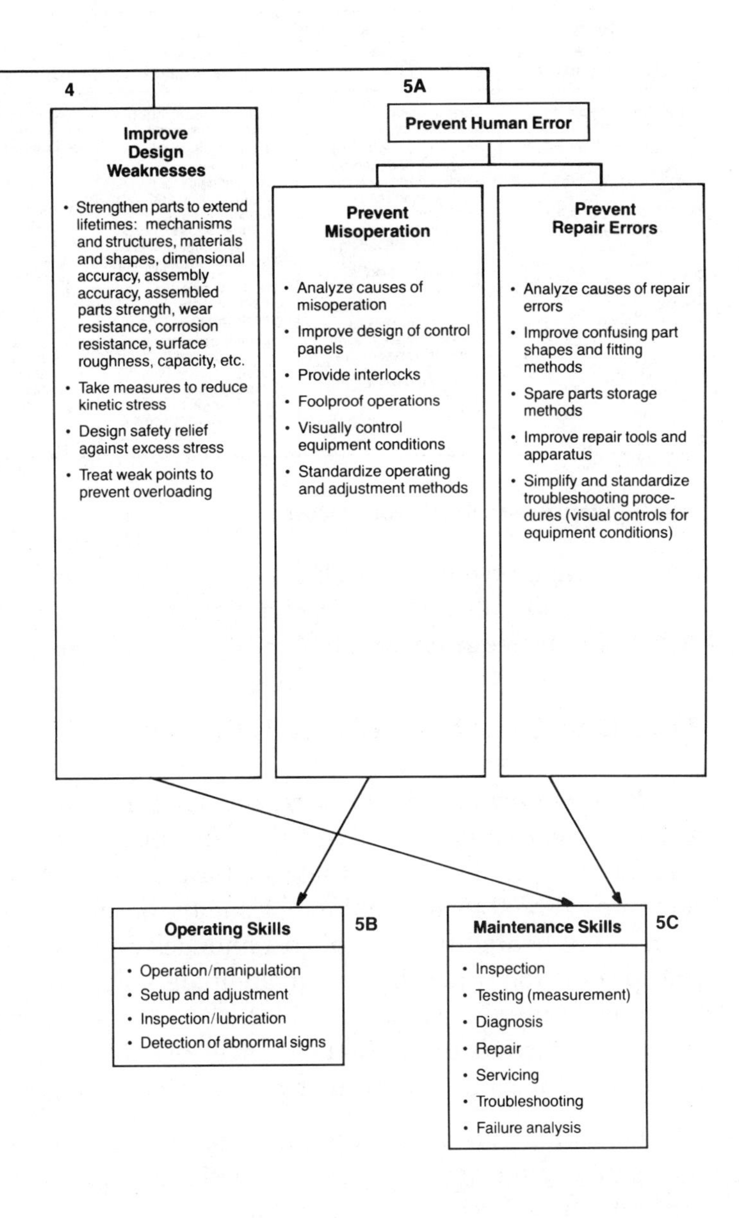
4
Improve Design Weaknesses
• Strengthen parts to extend lifetimes: mechanisms and structures, materials and shapes, dimensional accuracy, assembly accuracy, assembled parts strength, wear resistance, corrosion resistance, surface roughness, capacity, etc.
• Take measures to reduce kinetic stress
• Design safety relief against excess stress
• Treat weak points to prevent overloading
5A
Prevent Human Error
Prevent Misoperation
• Analyze causes of misoperation
• Improve design of control panels
• Provide interlocks
• Foolproof operations
• Visually control equipment conditions
• Standardize operating and adjustment methods
Prevent Repair Errors
• Analyze causes of repair errors
• Improve confusing part shapes and fitting methods
• Spare parts storage methods
• Improve repair tools and apparatus
• Simplify and standardize troubleshooting procedures (visual controls for equipment conditions)
Operating Skills
5B
• Operation/manipulation
• Setup and adjustment
• Inspection/lubrication
• Detection of abnormal signs
Maintenance Skills
5C
• Inspection
• Testing (measurement)
• Diagnosis
• Repair
• Servicing
• Troubleshooting
• Failure analysis

Phase 1	Phase 2	Phase 3	Phase 4
Stabilize Mean Time Between Failures (MBTF)	**Lengthen Equipment Life**	**Periodically Restore Deterioration**	**Predict Equipment Life**
Restore unchecked deterioration • Deal with visible defects **Prevent accelerated deterioration** • Set basic equipment conditions	**Correct design weaknesses** • Correct weaknesses in strength and precision • Select parts conformable to operating conditions • Correct weaknesses to prevent overloading **Eliminate sporadic breakdowns** • Improve operating and maintenance skills • Prevent misoperation • Prevent repair errors **Restore external appearance of equipment**	**Restore deterioration at regular intervals** • Estimate life span of equipment • Set standards for periodic inspection and testing • Set standards for periodic parts-replacement • Improve maintainability **Use the senses to detect internal deterioration** • Identify deterioration that gives warning signs • Identify types of warning signs given • Learn to detect warning signs	**Predict equipment life using diagnostic techniques** • Clarify and adhere to operating standards **Perform technical analysis of catastrophic failures** • Analyze rupture faces • Analyze material fatigue • Analyze gear tooth flanks, etc. • Take measures to extend equipment life • Conduct periodic restoration based on predicted life

Table 3-3. Zero Breakdowns in Four Phases

2. Lengthen equipment life.
3. Periodically restore deterioration.
4. Predict equipment life.

Phase 1: Stabilize Failure Intervals (MTBF)

Restore unchecked deterioration. First, restore equipment to its original condition. Generally, any maintenance worker asked to list problem areas that need immediate attention can come up with 30 to 40 problems. Usually these problems have been left untreated because of cost and lack of personnel, pressing production schedules, or inadequate engineering efforts. Carefully study the existing problems and set up an expedited treatment schedule to eliminate them, even if this means budgeting additional expenses and using subcontracted labor.

Prevent accelerated deterioration. Accelerated deterioration is the main reason for a large variation in equipment failure intervals. It is caused by poor maintenance of basic equipment conditions and neglect of operating standards — in other words, by people. Maintenance of basic equipment conditions and adherence to the operating standards will prevent accelerated deterioration and reduce the variability in equipment failure intervals.

Phase 2: Lengthen Equipment Life

Correct design weaknesses. If accelerated deterioration is checked, a piece of equipment (its component parts) will function for the length of its inherent life span, as determined by natural deterioration. The more a piece of equipment is limited to natural rather than accelerated deterioration, the smaller the variation in failure intervals and the longer its life. If equipment life remains too short in spite of the efforts mentioned above, there is probably a weakness in the design itself. Implementing improvements to ameliorate such weaknesses will lengthen equipment life. This strategy is generally referred to as "maintainability improvement" (increasing reliability).

Eliminate chance or accidental breakdowns. Although most chance breakdowns are caused by human operating errors, they can also be caused by repair errors. Moreover, chance failure in one part of the equipment often stresses other parts. Since this type of breakdown cannot be prevented by inspection or checks, operation and maintenance skills must be improved to eliminate the human errors that cause them.

At the same time, *poka-yoke* (mistake-proofing) devices should be used creatively at the sources of operating errors to stop them from generating defects or equipment breakdowns.

Restore visible deterioration. In phase 2 all visible external deterioration must be restored to original conditions. Generally, more than 50 percent of all breakdowns can be avoided by persistently restoring external deterioration.

Phase 3: Periodically Restore Deterioration

Estimate equipment life. Deterioration must be restored regularly to maintain the reduced level of breakdowns attained in phase 2, and to lower it even further. To do this, equipment life must be estimated as precisely as possible. Standards for periodic inspections, checkups, and part replacement must also be established and followed. At this stage thorough maintainability improvement is important. If standards are set without improved maintenance, the time, labor, and costs of analytic inspection and replacement of parts will balloon, and restoration will become impossible.

Learn signs of internal deterioration. Periodic restoration of external deterioration cannot prevent every breakdown. Operators must be trained to perceive the signs of abnormalities caused by internal deterioration. Although symptoms of internal deterioration are not always obvious, in many cases a trained operator can detect abnormalities in temperature, vibration, noise, light, color, smell, or movement. Operators and maintenance workers must analyze breakdowns thoroughly — on a daily basis, if necessary — to enhance their understanding of these abnormal signs. The following questions will facilitate this analysis:

- Were abnormal signs observed before the breakdown?
- Does this type of breakdown usually show abnormal signs?
- What types of signs typically precede such breakdowns?
- Why weren't the signs noticed in this case?
- How can the signs be noticed more easily?
- What additional knowledge and skill does the operator need to notice the breakdown signs?

Phase 4: Predict Equipment Life

Use machine diagnostic techniques. The actions described above will be very effective in preventing breakdowns and other losses in most equipment. In some equipment, however, life span remains unstable; breakdown signs cannot be detected by the five senses, or they are unreliable or too late. In such cases, machine diagnostic techniques can be used to detect otherwise invisible signs of incipient breakdown such as vibration, overheating, or precision problems.

Many types of diagnostic devices (both hardware and software) are available or are being developed. Superficial application is likely to be ineffective, but patient research should reward the user.

Analyze catastrophic breakdowns. Catastrophic breakdowns are completely unpredictable breakdowns causing total loss of all equipment functions. When breakdowns are reduced as a result of the four-phase program, only catastrophic breakdowns remain. If cost were not an issue, it might be possible to predict such catastrophic breakdowns, but in practice this is not reasonable.

Therefore, in the event of a catastrophic breakdown, a technical analysis of the underlying causes is still useful (*e.g.*, location of breakdown, equipment fatigue, mismatched gears, location of stress, and so on). With this knowledge, appropriate improvements can be made to maintain and lengthen the equipment's life.

Why the Four Phases Are Necessary

The four-phase program described above is effective in eliminating breakdowns for the following reasons:

From Accelerated Deterioration to Natural Deterioration

Breakdowns cannot be significantly reduced until the lives of component parts are lengthened and maintained in that condition at low cost (using periodic maintenance). This is accomplished in two steps: first, by eliminating accelerated deterioration so that only natural deterioration is influencing the equipment, and second, by correcting design weaknesses that shorten equipment life.

Accelerated deterioration must be eliminated first, for the following reasons:

- Accelerated deterioration must be eliminated so the potential life span, determined by natural deterioration, can be estimated.

- It is more economical to lengthen equipment life by eliminating accelerated deterioration than by attempting to lengthen the potential life span.
- Once accelerated deterioration is eliminated, equipment with a reasonably long life span does not need design improvements.
- Until accelerated deterioration is eliminated, true design weaknesses usually remain invisible.
- Even if equipment weaknesses are corrected, the overall effect is undermined or nullified if accelerated deterioration is still influencing the equipment.

Periodic Maintenance Begins with Longer, Stabler Life Spans

Economical, effective periodic maintenance begins with equipment life spans that have been stabilized and lengthened. When periodic maintenance is performed before the equipment life span is stable, maintenance costs are more expensive and the process is not effective. For example, consider a typical equipment life span distribution. Figure 3-8(a) shows a short life span spread over a wide range characteristic of accelerated deterioration. If periodic maintenance is applied under these circumstances, maintenance cycles will be short (thus expensive) and probably ineffective. The wider the range of equipment life spans, the harder it is to apply effective periodic maintenance.

Figure 3-8(b) illustrates the effect of phase 1: When life spans are stabilized and lengthened through the elimination of accelerated deterioration, the life span curve is closer to that determined by natural deterioration. Eliminating accelerated deterioration lengthens maintenance cycles (reducing maintenance costs) and reduces the likelihood of breakdowns.

Figure 3-8(c) shows the shift after phase 2, and the resulting change in life span distribution. As illustrated here, taking action against specific design weaknesses narrows the variation and extends life still further. This figure shows that switching to periodic maintenance in phase 3, after life span has been extended,

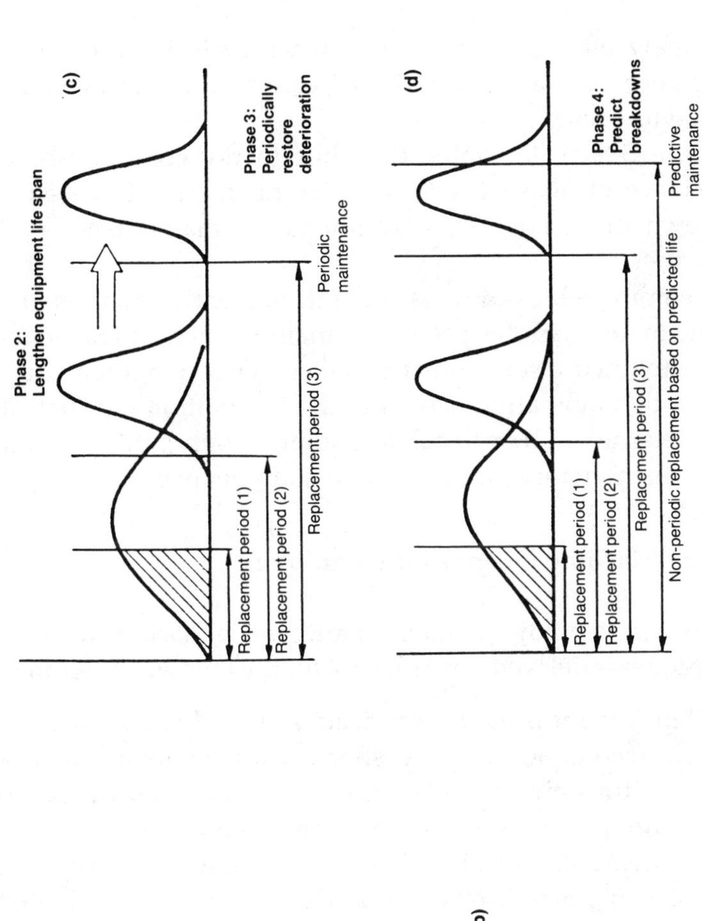

(a)
Breakdowns occur here
Replacement period (1)

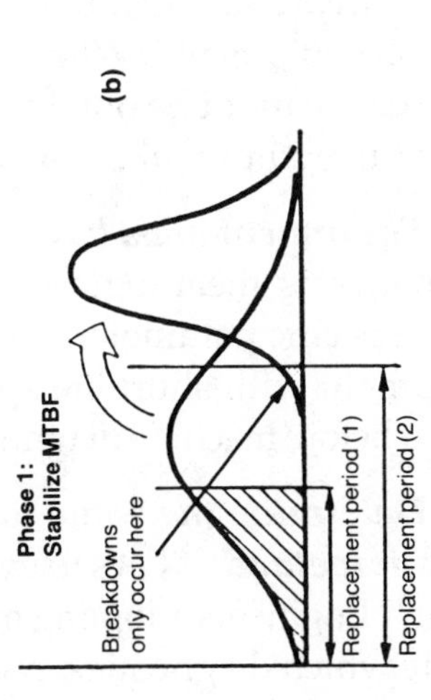

Figure 3-8. Four Phases Increase Maintenance Cycles and Lower Costs

is extremely effective. The maintenance cycle is even longer, maintenance costs are reduced further, and at each stage fewer breakdowns occur.

Figure 3-8(d) illustrates the shift to the condition-based maintenance of phase 4, or predictive maintenance, based on breakdown prediction. This lengthens the maintenance cycle even further.

To eliminate breakdowns through preventive maintenance (periodic maintenance + predictive maintenance) at reasonable cost, the following steps must be taken: First, lengthen equipment life by eliminating accelerated deterioration (so that life span is determined by natural deterioration only); then conduct periodic maintenance and predictive maintenance.

Applying Machine Diagnostic Techniques

The shift to predictive maintenance using machine diagnostic techniques is delayed until phase 4 for the following reasons:

*1. **Equipment must be serviced and calibrated.*** Machine diagnostic techniques employ shock pulse meters, vibration analyzers, ultrasonic detectors, and other techniques to measure deterioration parameters and perform quantitative analyses. The measured data must be reliable to be useful, however. This means the equipment itself must be serviced and well-calibrated. Reliable data cannot be obtained when accelerated deterioration is a factor in equipment behavior.

*2. **Equipment must have a reasonably long life.*** Predictive maintenance is meaningless unless deterioration is observed through specific parameters over a long period of time. The life of equipment with short life spans (less than one year) must be extended before machine diagnostic techniques will be successful.

*3. **Determine the value to be measured before selecting a diagnostic method.*** "Let's use a shock pulse meter — where can we use it?" Engineers adopting machine diagnosis techniques tend to decide which diagnostic device or method to use before deciding what to measure. A diagnostic method designed without a problem in mind is likely to be ineffective, however.

Each company should develop its own simple diagnostic methods, selecting devices from those available to suit its own unique needs. To use machine diagnostic techniques more effectively, first study the signs of abnormality in machinery and decide which characteristics of the machine should be measured. When the most useful parameters are clear, select appropriate devices or methods.

Case Study 3-1 — Four-Phase Development

Figure 3-9 shows the reduction in the number of breakdowns at Tokai Rubber Industries as each phase was introduced and developed. This company operates approximately 1,000 pieces of automatic equipment. Before improvement, the company experienced 1,000 breakdowns every month. After two years, at the end of phase 3, the rate had dropped to 90 cases per month. Eight months later, in phase 4, the number of breakdowns had been reduced to 20 cases per month.

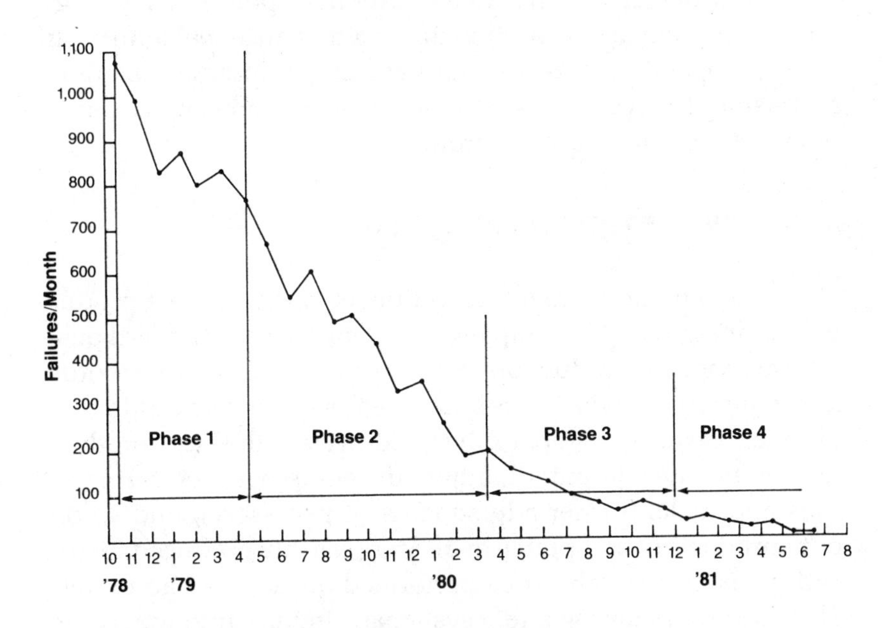

Figure 3-9. Reductions in Equipment Failures (Tokai Rubber Industries)

Maintenance Costs

Between phase 1 and phase 2 maintenance costs at Tokai Rubber Industries increased 10 percent over costs before the introduction of TPM. This increase funded efforts to lengthen equipment life. In phase 4, however, both maintenance costs and maintenance personnel were reduced by 15 percent. This company's experience demonstrates that while an initial investment is sometimes required, achieving zero breakdowns through preventive maintenance need not be expensive.

Distribution of Labor

Operators' involvement in autonomous maintenance is a key factor in implementing the four phases and eliminating breakdowns at reasonable cost. Operators' autonomous maintenance eliminates accelerated deterioration in phase 1 and sporadic breakdowns caused by operator errors in phase 2. Operators detect external deterioration in phase 2, although most treatment is performed by maintenance personnel. In phase 3, operators detect abnormal signs of internal deterioration using the five senses. Maintenance personnel carry out all other activities in phases 1 through 4.

IMPROVING SETUP AND ADJUSTMENT

Setup and adjustment downtime begins when the production of one product is completed and ends when standard quality is attained on production of the next product. In other words, it may include the time required to remove dies and jigs for one product, clean up, prepare dies and jigs for the next product, reassemble the equipment, adjust the equipment, perform trial runs and make further adjustments (if necessary), and so on, until products of acceptable quality are reliably produced. Setup and adjustment ought to be performed quickly and accurately. This requires planning and a systematic inquiry into ways of reducing setup and adjustment time without reducing accuracy.

Common Problems with Setup and Adjustment

With the widespread dissemination of Shigeo Shingo's *single-minute exchange of die* (SMED) techniques, and a growing reliance on an industrial engineering approach to factory problems, setup and adjustment times have been effectively reduced. [Editor's note: Because setup times of under 10 minutes are the standard in Japan today, this section assumes the reader has a basic familiarity with Shingo's techniques for reducing setup time. For a book-length treatment of this important subject, see Shigeo Shingo, *A Revolution in Manufacturing: The SMED System* (Cambridge: Productivity Press, 1985).]

Although setup times have been shortened, there is still room for improvement in many cases, particularly in the area of adjustments. For example, optimal reductions in setup time have been achieved when adjustments are simple, but less progress has been made on more complicated equipment.

The following discussion highlights some common problems.

Confused Procedures

People often complain about the length of setup and adjustment time, but few understand the relationship of the variables well enough to make real improvements. Uncertainty and inconsistency create obstacles to improvement in the following areas:

- working method (procedures, method, operator skills)
- jigs and tools (shape, mechanism, precision)
- precision (precision to be maintained, relationship between precision and adjustment)
- technical problems (technical improvements required)
- supervision (need for evaluation)

When these items are not treated and standardized and operators are left on their own, setup and adjustment times are inconsistent, and the causes remain uncertain.

Inconsistent Performance

Standardizing procedures can be difficult when equipment is operated inconsistently; if methods, procedures, and adjustments differ from worker to worker, setup and adjustment times fluctuate and troubles crop up later in production. Setup may have to be repeated if procedures have been standardized but not followed, or if appropriate procedures have not been standardized.

Adjustment Operations Unimproved

In general, adjustments account for 50 percent of setup time (Table 3-4). Often adjustments are not studied properly and problematic adjustments are left untreated.

Procedure	%
Preparation of materials, jigs, tools, and fittings	20%
Removal and attachment of jigs, tools, and dies	20%
Centering, dimensioning	10%
Trial processing, adjustment	50%

Table 3-4. Adjustments as a Percentage of Total Setup Time

Improving adjustment operations can reduce overall setup time considerably. First, unnecessary adjustments are eliminated. Then the time involved in performing unavoidable adjustments is reduced by various improvement strategies discussed in the next section.

Improving Setup

The first step in improving setup is distinguishing activities that can be performed while equipment is running from those that can be performed only when it is shut down.

The Difference Between Internal and External Setup

External setup activities are those that can take place while the equipment is running. They include preparing jigs, dies, and tools; preparing workbench and storage area for items to be removed; partial pre-assembly; and preheating. External setup activities can be performed in advance to save time when setting up the machine.

Internal setup activities can be performed only when the equipment is stopped, for example, when replacing dies and jigs, centering, and adjusting. Equipment downtime is reduced by eliminating from internal setup time all tasks that can be performed while equipment is running. This is the first step in setup improvement. (The terms internal and external setup were coined by Shigeo Shingo.)

Many external setup tasks are hidden in internal setup time. For example, when an essential tool or bolt is missing during the changeover operation, the operator must search for a substitute; when defects are found, time is taken to make partial repairs. Operators see these delays as normal occurrences, but an objective observer can see them as losses. The accumulation of small delays (1 to 2 minutes each, for example,) increases the overall time significantly. To eliminate these small losses, consider the following questions:

- What preparations need to be made in advance?
- What tools must be on hand?
- Are the jigs and tools to be installed in good repair?
- What type of workbench is needed?
- Where should jigs and dies be placed after removal? How will they be transported?
- What types of parts are necessary? How many are needed?

Simply considering these items in advance and making the necessary preparations substantially reduces setup time.

Seiri and *Seiton* in Setup Improvement

Three simple rules should be kept in mind in improving setup and adjustment:

- *Don't search* for parts or tools.
- *Don't move* unnecessarily; establish appropriate workbenches and storage areas.
- *Don't use* the wrong tools or the wrong parts.

These rules are related to the Japanese industrial housekeeping principles of *seiri* (organization) and *seiton* (tidiness), which are also important in setup improvement activities.* *Seiri* refers generally to the removal of unnecessary items (clearing) and the allocation of needed space (setting things in order) through workplace standards for placing, stacking, and storing. *Seiton* refers to the development of control techniques to ensure strict adherence to organizational standards (*e.g.*, visual controls).

In setup operations, these principles of organization determine what preparations are done while the machine is running to ensure that necessary items and tools will be available when and where needed, and in the right numbers. Operators apply the principle of tidiness in making sure the standards are followed, for example, developing checklists, designing visual controls on tool boards and parts containers, improving layout and position of workbench, and so on.

Separating Internal and External Setup

The first step in developing optimal setup operations is to study the current distribution of internal and external setup functions:

---

* *Seiri* and *seiton* (organization and tidiness) are two of the "Five S's" — organization or housekeeping concepts in Japanese industry. The cleanliness and orderliness of the best Japanese factories are the result of floor-level adherence to these concepts.

- Which tasks should be performed during external setup and what methods should be used?
- Which tasks should be performed during internal setup time and what procedures should be followed?

The following issues should be addressed for each task or unit of work:

1. Effectiveness of function served by the task
   - Is this task necessary? Can it be eliminated?
   - How redundant is it?
2. Effectiveness of work procedure (how task is performed)
   - What are the main points of the work procedure?
   - Are current procedures appropriate?
   - Is the procedure stable? (Can it be repeated consistently?)
   - How difficult is the procedure?
   - How can it be improved?
   - Can it be standardized?
3. Evaluation of work procedure
   - Is the current procedure optimal?
   - Should the order in which tasks are performed be changed?
   - Can some tasks be combined?
   - Can any functions be performed simultaneously?
4. Distribution of tasks (division of labor)
   - Is the distribution of tasks optimal?
   - Is the number of personnel adequate?

Implementing improvements discovered through this kind of inquiry can reduce setup time by 30 to 50 percent.

To minimize losses attributable to setup and adjustment, make certain that no problems with quality occur after restart and that procedures are standardized so everyone can perform them in the same amount of time. All mechanical, procedural, and technical problems should be well understood, and areas needing further study should be clearly identified.

Converting Internal to External Setup

Many functions currently performed during internal setup can be performed during external setup time, or modified to reduce their duration, using an analysis similar to one in the previous section. For example, a jig that is typically changed, assembled, and adjusted while the machine is stopped can be preassembled while the machine is still running. Adjustments that must be made during internal setup time can be reduced through partial adjustment during external setup time according to predetermined standard procedures.

The following methods can be used to convert internal to external setup:

1. ***Preassemble.*** Rather than attaching each part separately during internal setup, preassemble the parts during external setup and then position the assembly during internal setup.

2. ***Use standard and "one-touch" jigs.*** Compare the shapes of the tools and jigs for different products and consider preparing a standard jig that can be shared by all. Consider jigs that quickly position workpieces without the operator's help (quick-fitting or "one-touch" jigs).

3. ***Eliminate adjustments.*** Wherever possible, avoid making adjustments during internal setup time. For example, eliminate the need for adjustment by establishing constant values that can be set in one touch without trial-and-error adjustment.

4. ***Use intermediary jigs.*** Whenever a cutting tool is changed, it must be centered and adjusted. By presetting the tool on a precise and standard intermediary jig, the tool and device can be attached directly to the base, eliminating the need for centering.

Shortening Internal Setup Time

Internal setup time itself can often be reduced by using quick-fitting jigs and improved assembly and fastening methods, and by eliminating adjustments.

*1. **Simplify clamping mechanism.*** Reduce the number of bolts, use more efficient clamping devices, or replace bolts with an oil hydraulic or cam-operated clamp to attach jigs to machinery.

2. Adopt parallel operations. Two people working together simultaneously can perform a setup faster and more effectively than a single person performing each step in sequence. Although timing and coordination are crucial in such operations, in the best cases, setup times can be halved with the same number of work hours.

3. Optimize the number of workers and division of labor. Large, complicated setups must sometimes be performed by dozens of people. In such cases, considering the following points can reduce setup and adjustment time drastically:

- What is the optimal number of workers for each task?
- How should the work be divided or shared?
- What are the critical paths? Can they be reduced?
- How can manpower be used more effectively?

Table 3-5 is an example of strategies for improving setup.

Eliminating Adjustments

Many adjustments can be performed without trial-and-error experimentation. Only unavoidable, essential adjustments should remain.

To eliminate adjustments, analyze their purpose, causes, actual methods employed, and effectiveness.

Purpose of Adjustment

Adjustment accomplishes the following basic purposes:

- *Positioning*: setting position on X, Y, or Z axes
- *Centering*: centering cutting tools on workpieces, for example

External Setup	Preparations	• Tools (types, quantities) • Locations • Position • Workplace organization and housekeeping • Preparation procedure	• Don't search • Don't move • Don't use
	Preparation of ancillary equipment	• Check jigs • Measuring instruments • Preheating dies • Presetting • Internal Setup	
Internal Setup	Operation phase	• Standardize work procedures and methods • Allocate work • Evaluate effectiveness of work • Parallel operations • Simplify work • Number of personnel • Simplify assembly • Assembly/integration • Elimination	• Eliminate redundant procedures • Inculcate basic operations
	Dies and jigs	• Clamping methods • Reduce number of clamping parts • Shapes of dies and jigs — consider mechanisms • Use intermediary jigs • Standardize dies and jigs • Use common dies and jigs • Weight • Separate functions and methods • Interchangeability	
	Adjustment	• Precision of jigs • Precision of equipment • Set reference surfaces • Measurement methods • Simplification methods • Standardize adjustment procedures • Quantification • Selection • Standardization • Use gauges • Separate out interdependent adjustments • Optimize conditions	Eliminate adjustment

Table 3-5. Improving Setup and Adjustment

- *Measuring*: adjusting the cutting depth to design dimensions, for example
- *Timing*: adjusting timing on various equipment functions
- *Balance*: adjusting pressure, balance of springs, or balancing with setting screws

Causes of Adjustment

Adjustments are needed in the following circumstances:

Accumulation of errors. When precision in equipment is poor and when there are no specific control limits to maintain, the effects of small errors begin to multiply and must be corrected periodically through adjustment. Furthermore, imprecise adjustment of jigs and tools often combines with lack of equipment precision to magnify adjustment problems. Correcting accumulated errors accounts for a large percentage of total adjustment time.

For example, equipment is often provided with fitted shims that correct the effect of abrasion on guides. This procedure should be avoided, however, because while easy to perform, it is ultimately time-consuming. Clearly defined control limits and maintenance of precision will prevent the effect of accumulated deviations and eliminate altogether the need for this type of adjustment.

Lack of rigidity. If everything checks out when the machine isn't running, but errors are produced during operation, the equipment or parts may be flexing.

Lack of standards. Adjustments are required when there are no standards, when standards are inadequate, or when not enough data is available to permit standards to be set. Adjustment is always necessary when no datum plane has been specified. Even when reference marks are available, they will not be useful if there is surface roughness or oil buildup, for example.

Lack of measuring method. Adjustments are required when measuring methods are not available or when inability to quantify prevents a method from being established.

Measuring methods may be inadequate because the company is unaware of appropriate methods already used at other companies. In such a case, existing measurement methods and devices should be researched and adopted. Sometimes, however, appropriate methods have simply not been developed. In those cases, an appropriate method must be developed in-house. Adjustments may or may not be necessary.

Unavoidable adjustments. Some types of equipment mechanisms require human intervention and adjustment for proper functioning. Unless the mechanism is redesigned, adjustment is required.

Improper work methods. Some adjustments are necessary because work methods and procedures are unclear. Even when they are clear, they may be performed incorrectly because the results are not checked adequately. Such cases can easily be avoided.

Analyzing Effectiveness of Adjustment Operations

Use the following analysis to study the effectiveness of adjustment operations and to determine which adjustments are essential and which can be eliminated (Table 3-6 and Figure 3-10).

Purpose What function is apparently served by adjustment?
Current Rationale Why is adjustment needed at present?
Method How is the adjustment performed?
Principles What is the true function of the adjustment operation as a whole?
Causal Factors What conditions create the need for adjustment?
Alternatives What improvements will eliminate the need for adjustment?

Table 3-6. Conceptual Steps for Analyzing Adjustment Operations

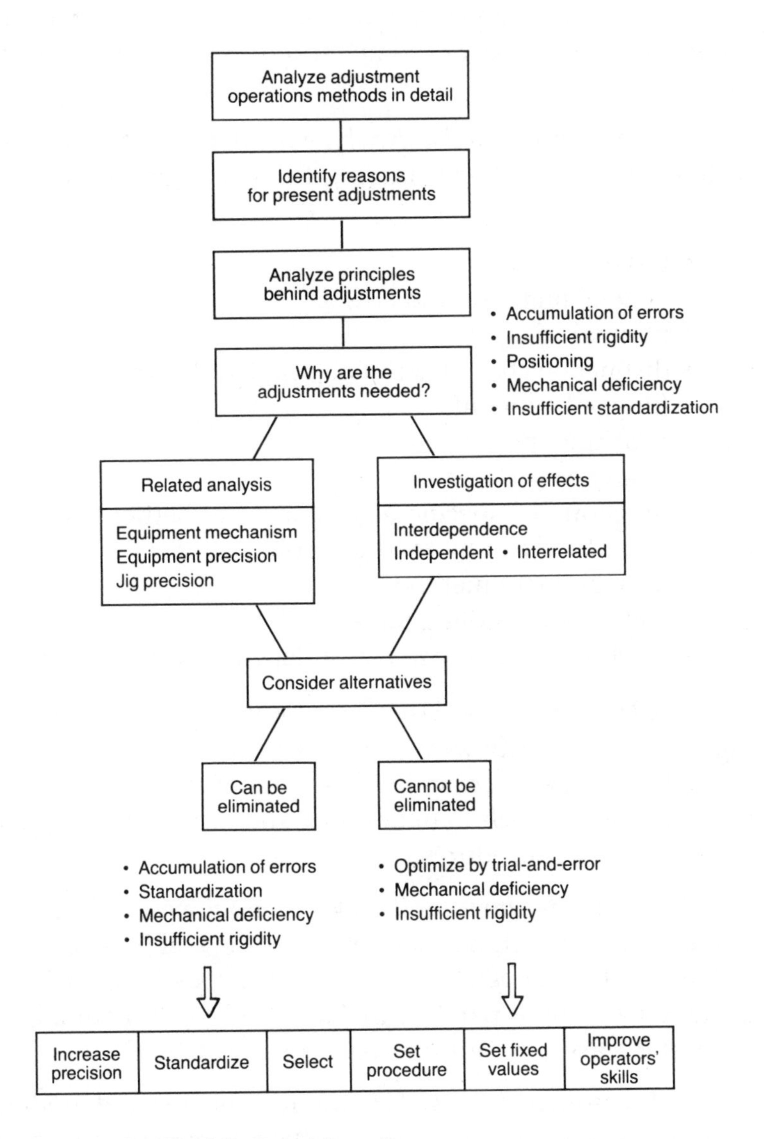

Figure 3-10. Analysis of Adjustment Operations

a. Identify purposes. Identify the apparent purpose of each adjustment (centering, positioning, measuring, etc.), paying particular attention when the purpose served by the adjustment is unclear. Note that some procedures will have more than one purpose.

b. Analyze methods. Analyze the details of the current adjustment operations, considering the following points, where appropriate:

- order of steps
- general methods and criteria
- number of repetitions
- distinctions between initial (rough) adjustments and final (fine) adjustments
- attachment methods
- reference surfaces
- functions of specific adjustments (whether each operation has one or more functions)
- measurement methods
- method of handling materials
- whether adjusted parts are independent or interconnected

c. Clarify reasons. Use the insights gained in the preceding analysis to clarify why each procedure is currently needed. Consider the operations individually and in groups, and investigate the aims of each in detail. List and organize the apparent reasons for the procedures.

d. Analyze principles. Look beyond the procedures again, to the principles behind them. What is the real function served by the adjustment operation? For example, does it balance heights, make left and right parallel, make something horizontal, align the X and Y axes, or perform some other function?

e. Investigate causes. Use the results of step *d* to identify why the adjustment is necessary. Is the adjustment caused by an accumulation of errors, lack of rigidity, insufficient standardization, mechanical deficiency, or something else? The adjustment may be caused by one or more sets of circumstances.

f. Consider alternatives. Finally, consider improvements that will eliminate the need for any of the adjustments.

Improving Unavoidable Adjustments

When adjustments cannot be eliminated, several strategies to streamline them may be adopted.

Set Fixed Values

Use constant value settings to avoid adjustments wherever possible. Otherwise, consider measuring methods that might make numerical values possible, or try using a different attribute for determining the adjustment.

Establish a Procedure

Establish a standard procedure for performing the adjustment and make sure each step is thoroughly understood. After each step the worker should ascertain that the correct result was produced and that the adjustment was within the correct range before going on to the next step.

Also consider how adjusting one part of a machine affects other parts. Find ways to minimize an adjustment's effects on other quality characteristics and to facilitate subsequent adjustments.

Improve Skills

To avoid errors, increase workers' skills by having them practice the procedures. Skills perfected through repetition are retained over a long period.

Steps for Improving Setup and Adjustment Operations

Figure 3-11 is a systematic overview of the program for improving setup and adjustment operations.

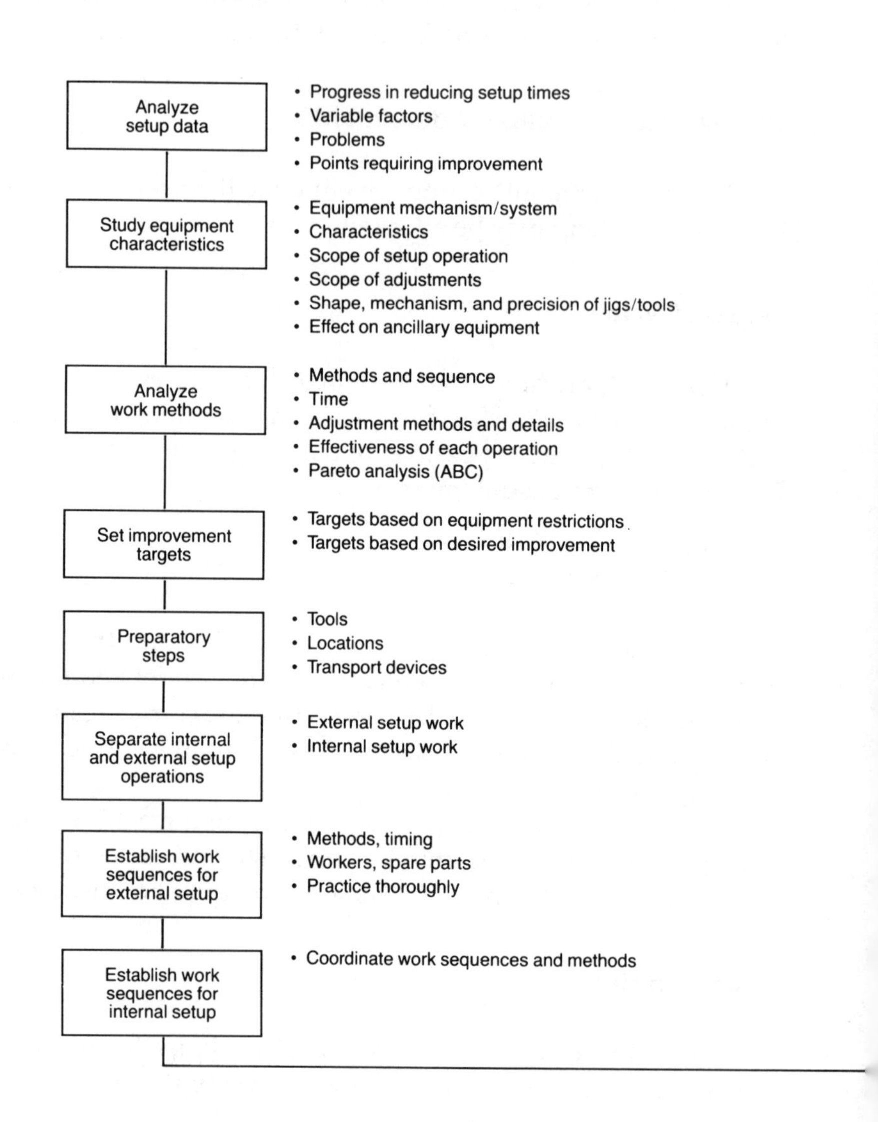

Figure 3-11. Improving Setup and Adjustment

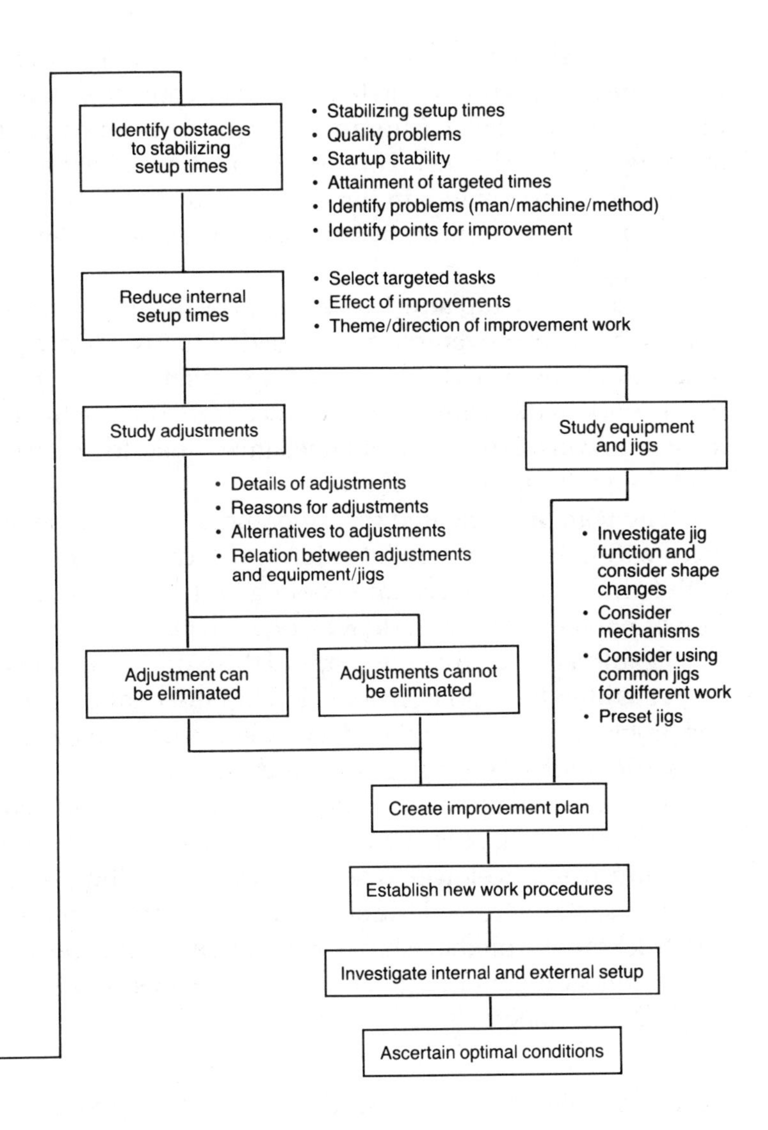
Identify obstacles to stabilizing setup times
• Stabilizing setup times
• Quality problems
• Startup stability
• Attainment of targeted times
• Identify problems (man/machine/method)
• Identify points for improvement
Reduce internal setup times
• Select targeted tasks
• Effect of improvements
• Theme/direction of improvement work
Study adjustments
• Details of adjustments
• Reasons for adjustments
• Alternatives to adjustments
• Relation between adjustments and equipment/jigs
Study equipment and jigs
• Investigate jig function and consider shape changes
• Consider mechanisms
• Consider using common jigs for different work
• Preset jigs
Adjustment can be eliminated
Adjustments cannot be eliminated
Create improvement plan
Establish new work procedures
Investigate internal and external setup
Ascertain optimal conditions

Case Study 3-2 — Eliminating Adjustment

In a flash butt-welding process the working faces of the upper and lower chucks had to make close contact with the workpiece to prevent level differences, gaps, uneven contact, and slipping in the chucks. This adjustment was carried out by clamping the workpiece in the chucks and checking its contact, then filing the chuck faces by hand and inserting a shim in one of the chuck holders.

Most of the setup time was taken up by this adjustment operation, and the improvement team decided to investigate possible time reductions. The chucks were roughly set to match the shape of the work in the external setup procedure, but the final adjustment was considered unavoidable since it had to be carried out with the workpiece in position.

The team investigated the differences in the shape of workpiece within lots and from lot to lot. It also checked the precision of the chucks (die blocks and cooling chucks), including precision of parallel edges and dimensional precision.

Results showed that the shape of the work varied to a certain degree within lots and from lot to lot, but not enough to be a serious problem. The team did find a problem related to the precision of the die blocks, cooling chucks, and other parts.

The adjustment was needed to correct the accumulated errors in the parallel edges of the chucks and die blocks, both left-to-right and upper-to-lower. The team increased the precision of the blocks, standardized the shims, and conformed the shape of the chucks to the median shape of the work. With these improvements, the adjustment operation was no longer necessary (Figure 3-12 and Table 3-7).

Case Study 3-3 — Improving Unavoidable Adjustments

This example concerns a cold forging plant for automobile screws. Adjustments were extremely difficult, since the various adjustment procedures were interdependent and had complex effects on the quality characteristics. There was no problem when adjustment work went well, but when it did not, which was often, adjustments took several hours. The improvement team adopted the following measures to remedy this problem (Table 3-8): First, the team identified the conditions under which the adjustment went smoothly. Then, it quantified as many of these conditions as possible and established standard settings. The optimal adjustment procedure was firmly established, and the standard settings for conditions were always used in setting up.

These improvements greatly reduced the number of trial runs (as shown in Figure 3-13), cut the overall setup time drastically, and reduced the variation in setup times.

REDUCING IDLING AND MINOR STOPPAGES

Idling and minor stoppages occur when equipment idles (*i.e.*, continues to run without producing) or stops as a result of a temporary problem. For example, a minor stoppage occurs when a workpiece is jammed in a chute or caught on an obstruction (cutting off the supply), or when a sensor activates and shuts down the machinery. These troubles are usually noticed quickly; normal operation can be restored by simple measures such as removing or reinserting the jammed workpiece correctly or by switching the equipment back on.

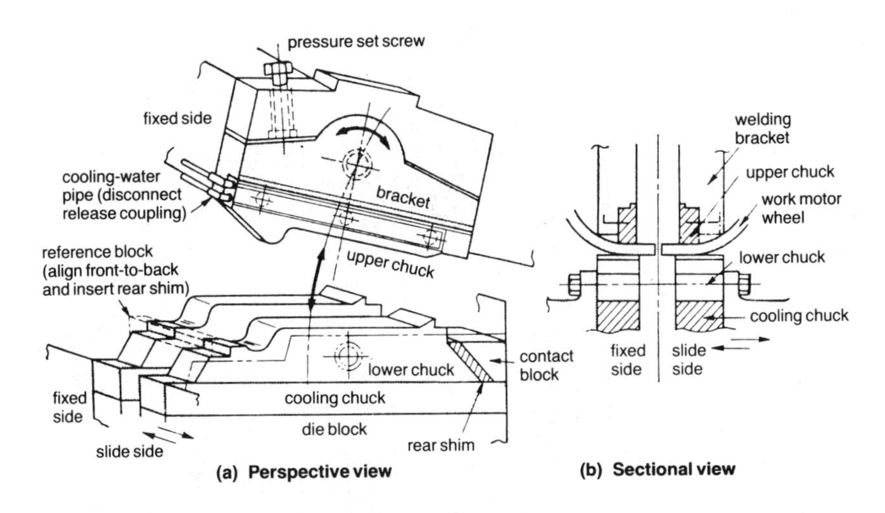

Figure 3-12. Upper and Lower Chucks in Flash Butt-Welding Process

Purpose	Rationale	Method	Principles	Causal Factors	Precision of relevant parts	Alternatives
• File working faces of chucks to fit work	• Adjust chuck faces to variation in shape of work within lots and from lot to lot • Check mounting of upper and lower chuck and set shim • Rotate turnbuckle to distribute clamp pressure evenly over work and ensure uniform contact of upper and lower chuck faces with work	Omitted	• Horizontal and vertical parallelism, horizontality, and displacement	• Lack of standardization • Accumulation of errors • Lack of standardization	• Precision of upper faces of left and right cooling chucks: 0.6–0.8mm • 1–2mm difference between left and right in relation to rear contact block • Because chucks are re-used, 10–15mm difference compared with new chuck	Can be eliminated

Table 3-7. Analysis of Effectiveness of Adjustment

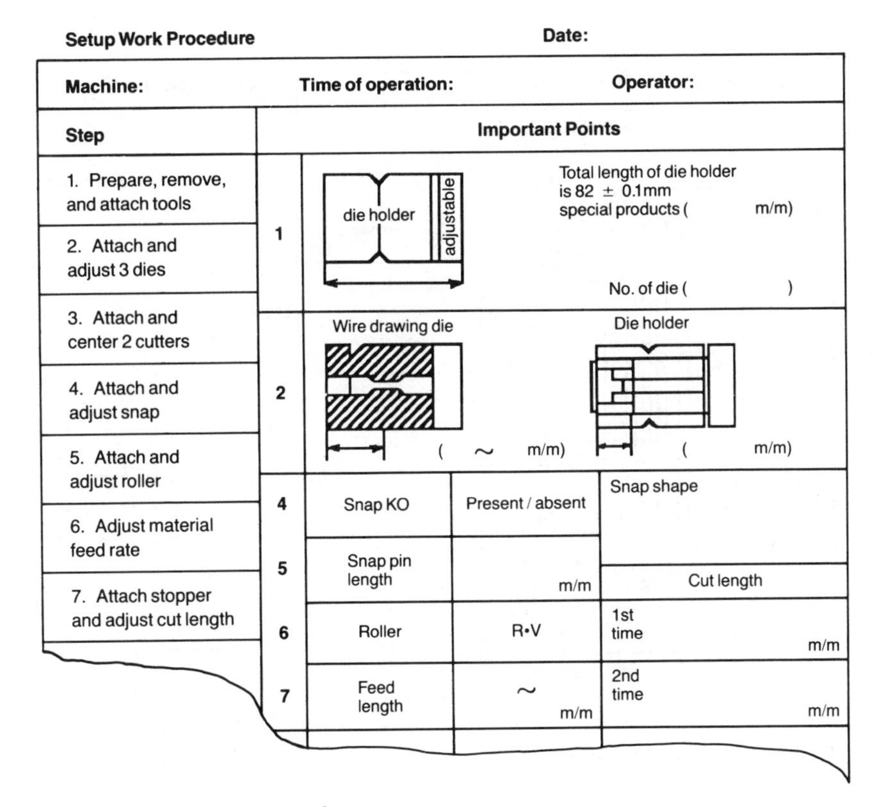

Setup Work Procedure **Date:**

Machine:	Time of operation:	Operator:

Step
1. Prepare, remove, and attach tools
2. Attach and adjust 3 dies
3. Attach and center 2 cutters
4. Attach and adjust snap
5. Attach and adjust roller
6. Adjust material feed rate
7. Attach stopper and adjust cut length

Important Points

1

Total length of die holder is 82 ± 0.1mm
special products (m/m)
No. of die ( )

2
Wire drawing die (~ m/m)
Die holder (m/m)

4	Snap KO	Present / absent	Snap shape
5	Snap pin length	m/m	Cut length
6	Roller	R•V	1st time m/m
7	Feed length	~ m/m	2nd time m/m

Table 3-8. Setup Work Procedure

Since breakdowns cause a loss or reduction in normal equipment functions, restoring normal conditions requires repair work, parts replacement, and adjustment. This takes time. Since idling and minor stoppages interrupt functions, they can also be categorized as breakdowns. Even so, the two are essentially different. A minor stoppage can be dealt with quickly, as soon as it is noticed. On the other hand, if the equipment is idle or stopped frequently, the factory's output may drop. This occurs more frequently in factories with a large number of automatic machines. If not detected quickly, a minor stoppage soon turns into a major stoppage and a major cause of reduced operating rates.

In factories with many automated production systems, idling, minor stoppages, and associated defects prevent individual

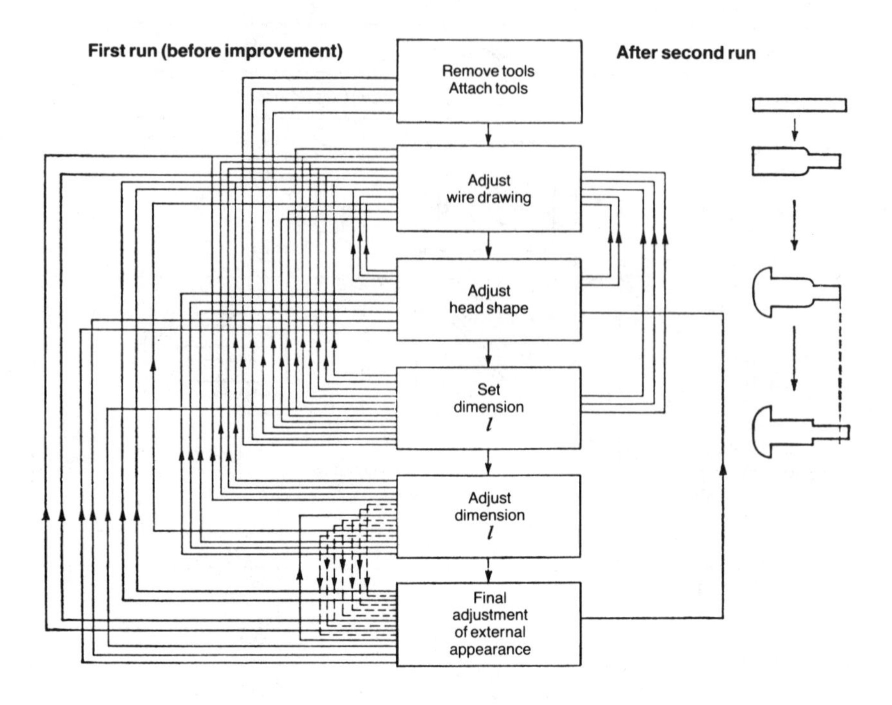

Figure 3-13. Test Runs during Setup

machines from being used to their full capacity. Operators try to keep automated machinery running smoothly so as not to waste its capacity, but they are also kept busy coping with problems. Since their efforts are too diffused to produce any improvement, the machinery is automated in name only. We might say that the operators, rather than using the machines, are used by them.

Many companies aiming at complete automation have set up unmanned plants. In many of these, however, even a single minor stoppage will bring the factory to a halt, halving the effect of the automation and reducing the operating rates.

Definitions

Stoppages occur when a problem is detected by an instrument and the equipment stops automatically.

Stoppages due to overloading. Some stoppages are caused by overloading, often found in automatic packers and assemblers. They occurs when workpieces collide, for example.

Stoppages due to quality abnormalities. These occur in automatic assemblers, other automatic machinery, and transfer machines. For example, sensors trip and stop equipment automatically when assemblers fail to pick up parts correctly and misassembly occurs.

Idling. Idling occurs when the flow of workpieces stops but the equipment continues to run without processing. If the mechanism of the equipment prevents this from being observed, or if sensors are too costly to install, idling may not be noticed for some time. Idling occurs in all kinds of automatic machinery when there are defects in the mechanisms that feed or transport the work.

Characteristics of Idling and Minor Stoppages

A number of characteristics of idling and minor stoppages make them difficult to address systematically.

Easy to Restore

Idling and minor stoppages are easily restored, so little effort is made to eradicate them. Typically, production and maintenance personnel do not regard idling and minor stoppages as problems but simply tolerate them.

Conditions of Occurrence Vary Widely

Idling and minor stoppages may occur with certain products or parts but not others, or with all products or parts but only under certain conditions. They may occur only on certain days, or only with certain machines. These varying conditions inevitably make them easy to ignore (Figure 3-14).

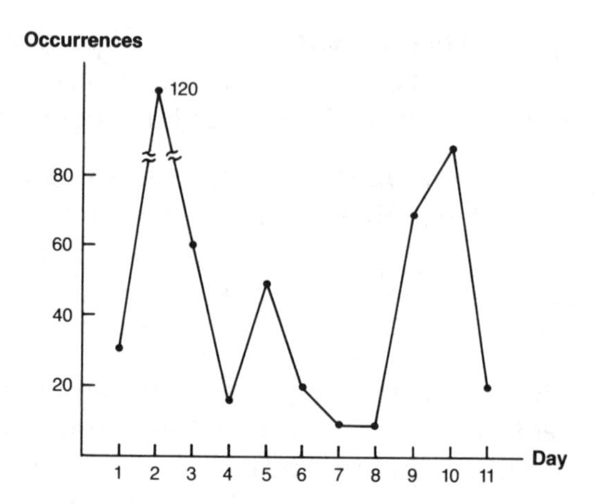

Figure 3-14. Variation in Idling and Minor Stoppages with Same Product

Location Constantly Changes

Idling and stoppages seldom occur at the same location in the machine. Often an outbreak in one area is followed by separate outbreaks in different areas, making their true nature difficult to grasp; the problem may be chronic, or a sporadic problem may occur along with a chronic problem.

In the first case, the idling and minor stoppages are concentrated in a certain portion of the machine. Various actions reduce their occurrence; then they break out at another location. As a result, there is no overall improvement. The concentration of occurrences in a particular location focuses attention exclusively on that area while the real problem — a hidden defect — is located elsewhere in the machine. Since it is invisible, that hidden cause is overlooked.

Thus, a key strategy in reducing idling and minor stoppages is the single-minded pursuit of hidden defects. Improvement teams must be determined to eradicate them wherever they might occur — regardless of their frequency.

In the second case, a sporadic, unexpected stoppage occurs simultaneously with a chronic stoppage. The sporadic stoppage may be caused by a faulty part or an incorrectly installed jig; but whatever its cause, the sporadic problem is a different phenomenon from the chronic problem and manifests itself differently. It is therefore important to notice that difference quickly and take corrective measures.

Scope of the Loss Unclear

The losses caused by idling and stoppages are difficult to quantify over long periods. Such factors as location, frequency, time required for correction, and so on, are difficult to measure continuously, although longer periods might be tracked if each operator were responsible for a single piece of equipment. Operators in charge of large numbers of machines, however, can only determine the net operating time and estimate the idling and minor stoppage time losses from the production figures. Unfortunately, this gives no information on the number of occurrences.

Common Problems

Typically, the losses due to idling and minor stoppages are not adequately measured, treated, or observed.

Losses Remain Unnoticed

Even though a minor stoppage is easily corrected each time it happens, the losses in production resulting from frequent occurrences or from stoppages that are not discovered quickly are surprisingly high. Since the losses are unnoticed and unmeasured, however, the problem is not acknowledged. Thus, a preliminary step in solving the problems of idling and minor stoppages is to measure the losses they cause.

Remedial Actions Are Inadequate

Typically, operators and maintenance personnel treat idling and minor stoppages superficially, applying stopgap measures or remedies that deal with only part of the problem. They treat the symptoms but do not take the fundamental measures needed to eradicate the causes.

The Phenomena Are Not Observed Closely Enough

Seeing what actually takes place when idling and stoppages occur is an important key to their solution. Being in the right place at the right time is difficult, however, and the phenomenon may be too brief to be clearly observed. This contributes to the tendency to treat results rather than causes.

To attack the basic causes of idling and minor stoppages, an improvement team should observe them happening on the shop floor and then analyze and classify the results. The roots of the problems must be identified and understood before remedial action is taken.

Strategies for Reducing Idling and Minor Stoppages

This section reviews basic TPM strategies for reducing idling and minor stoppages.

Correct Slight Defects in Parts and Jigs

To begin reducing idling and minor stoppages, search out and correct all the slight defects in parts and jigs involved in the transfer of work. These slight defects often consist of minute irregularities in the external appearance and shape of parts. Use magnification and scientific observation methods that enhance the power of the five senses. Analytical precision is important in detecting the small differences leading to these problems, and sometimes requires the development of new measuring methods. Evaluation criteria are particularly important for unmeasurable phenomena since these can only be judged through the five senses.

The typical approach to equipment problems can be a major obstacle to improvement. We cannot unearth the roots of chronic problems and find new solutions until we are able to see details we never noticed before and to consider innovative solutions. Use the following strategies to change the way groups think about equipment problems:

- Acknowledge that problems are there to be found, in other words, that room for improvement exists.
- Uncover hidden problems by comparing things as they are with how they should be.
- Investigate *anything* out of the ordinary — anything that might be a potential factor — in accordance with sound engineering theories and principles.

Correcting slight defects minimizes the variation in location and frequency of idling and minor stoppages from lot to lot and day to day. In other words, it

- narrows down the causes of idling and minor stoppages
- reveals differences in the occurrence of idling and minor stoppages before and after the correction of slight defects
- brings hidden defects to light

Table 3-9 shows the results of careful investigative analysis of minor stoppages and idling. Table 3-10 organizes various approaches to improvement that ensure systematic and thorough investigation.

Ensure that Basic Equipment Conditions Are Maintained

Idling and minor stoppages are frequently caused by failure to maintain basic equipment conditions (cleaning, lubrication, bolting), so make certain these standards are scrupulously observed. If equipment is left dirty because no one has bothered to clean it, or if play and looseness are not corrected, stoppages are inevitable. It is vital that workers understand and maintain the basic conditions of their equipment.

Type	Cause	Type	Cause
Transfer Line 1. Clogging 2. Catching 3. Jamming 4. Overlapping 5. Parts running out 6. Insufficient feed 7. Excess feed 8. Falling off 9. Mis-insertion	1. Causes originating in materials and parts • Defective dimension • Defective exterior or shape • Inclusion of wrong type of material or part • Presence or absence of magnetic field 2. Causes originating in transfer or feed systems • Defective chute (incorrect shape, poor surface condition, damage, dirt, uneven connecting parts) • Defective parts feeder (vibration amplitude, resonance, imbalance, non-optimal feed rate, misattachment) • Defective positioning control (poor method, unsuitability for parts, non-optimal feed rate)	**Assembly System** 1. Crushing or other damage 2. Double feed 3. Misinsertion in chuck 4. Timing 5. Defective assembly 6. Mis-ejection	3. Cause originating in assembly system • Jig precision • Assembly • Parts precision • Timing 4. Causes originating in production management • Adjustment error during setup • Incorrect setting
		Detection System 1. Faulty response	5. Causes originating in detection system • Defect in system itself • Poorly-installed sensor — wrong method or position • Wrong sensitivity setting • Incorrect adjustment • Timing • Unsuitable operating conditions

Table 3-9. Types of Idling and Minor Stoppages and Their Causes

	Approach to Improvement	Object of Improvement
Reliability of use	1. Correct minor defects	a. External appearance (surface damage, wear) b. Dimensions (necessary dimensional precision, clearances) c. Actuation (play, eccentricity)
Reliability of use	2. Apply basic principles of shop-floor operations	a. Cleaning (dirt, play) b. Lubrication (dirt, wear) c. Bolting (loose bolts and nuts)
Reliability of use	3. Adhere to basic work procedures and standards	a. Correct manipulation b. Setup (adjustment methods, settings) c. Observation of equipment (methods of detecting abnormalities)
Reliability of use and equipment fabrication	4. Identify optimal conditions	a. Installation conditions (angle, position, resonance, compressed-air pressure, degree of vacuum, vibration amplitude, etc.) b. Processing conditions (optimum feed rate, etc.)
Reliability of use and equipment fabrication	5. Identify required configuration	a. Limits of required precision (parts precision, assembly precision) b. Conditions of use (optimal range of use)
Inherent reliability	6. Investigate design weaknesses	a. Designs conformable to shape of parts (shape-design change) b. Selection of parts (change resulting from material quality/function) c. Consideration of mechanisms and systems

Table 3-10. Six Approaches to Improving Idling and Minor Stoppages

Review Basic Operations

The frequency of idling and minor stoppages is often affected by the way equipment is set up or adjustments are carried out. The same operator may produce different results on different days depending on how he or she set up the equipment.

Make certain that setup, adjustments, and other operations are being carried out correctly. Even if basic operations have been taught, it pays to inspect periodically for proper performance. A thorough review of procedures may be necessary to prevent problems caused when basic operations are carried out incorrectly or not at all.

Conduct Physical Analysis of Phenomena (P-M Analysis)

Thoroughly implementing the three strategies described above (correcting slight defects and maintaining basic equipment and operating conditions) alters the occurrence, frequency, and location of idling and minor stoppages. These improvements alone cannot reduce problems to zero, however. To reduce their occurrence still further, they must be analyzed according to physical principles, using P-M analysis (*see* Chapter 2).

For example, an assembly shop at factory C experienced production losses of 40 to 50 percent as a result of idling and minor stoppages. The assembly machine was installed on a six-station index table and the parts were supplied automatically. In the next process, they passed through parts feeders and were assembled by a vacuum gripper. The assembly was then inspected automatically, and final assembly was performed in the last step.

The sensor was set to shut down the equipment if a part was not picked up by a suction nozzle. Idling and minor stoppages occurred when the nozzles failed to pick up parts and when the sensor did not operate correctly. The first problem is treated in this example of P-M analysis.

As shown in Table 3-11, a reduction in suction pressure was the physical cause of the nozzles' failure to pick up parts. When a temporary suction drop, for whatever reason, prevents the nozzle from picking up a part, a sensor shuts down the equipment.

Adopt an Analytical Approach

To conduct a P-M analysis in this case, consider the conditions needed to produce a drop in suction pressure. Take a systematic approach to identify all the conditions that would be certain to produce such a drop. If the analysis is not well organized, the main cause of the problem is likely to remain hidden. The most important strategy at this point is to avoid reaching for what appears at first to be the most obvious cause.

Phenomenon	Physical Characteristics	Conditions for Occurrence	Relation to Work and Equipment
Stopped in response to sensor Poor vacuum nozzle suction	Loss of suction	1. Workpiece deformed	a. Change in shape b. Changed dimensions
		2. Faulty actuation of vacuum system	a. Degree of vacuum too low b. Degree of vacuum varies c. Faulty timing
		3. Intake of air on contact face	a. Wear/diameter of vacuum nozzle b. Wear of work supply jig c. Faulty queuing of work at supply jig d. Poor contact
		4. Eccentricity	a. Suction nozzle and supply jig off-center b. Play of suction nozzle c. Displacement by vibration (due to resonance)

Table 3-11. Analyzing Idling and Minor Stoppages

Analyze each possible set of conditions in relation to the machines, tools, jigs, and materials, without omitting any factor. Then, think through the causes producing each set of conditions and determine what would happen if the causes were to change.

For example, poor contact between a suction nozzle and the workpiece might cause air to be taken in, reducing the suction. Contributory causes might be nozzle wear, a worn workpiece supply jig, misalignment of the workpiece with the supply jig, and irregularities on the workpiece surface in contact with the nozzle. These factors, singly or in combination, could certainly cause the intake of air at the suction nozzle.

After this kind of analysis, consider each factor, listing unsatisfactory points and small telltale signs of trouble, and then plan appropriate corrective actions. When this systematic approach is used, the actions taken are certain to affect the problem.

Determine Optimum Conditions

The next strategy for reducing idling and stoppages is addressing equipment reliability. Taking the existing equipment, jigs, and tools as the starting point, review the installation and processing conditions of all parts and units and consider how these conditions can be optimized. Installation conditions include every factor related to the way the equipment is installed, such as position of parts, angles, and resonance. Processing conditions are the physical properties relating to processing, such as pneumatic pressure, vacuum pressure, vibration amplitude, and workpiece supply volume.

Installation and processing conditions are often established based on past experience. Whether or not they are optimal is a separate question that is not always addressed. For this reason an experimental, trial-and-error approach must be adopted in reviewing the existing conditions.

Eliminate Design Weaknesses

If the preceding approaches do not reduce stoppages, the root of the problem is often design weaknesses in equipment, jigs, tools, or detection systems. Weaknesses and possible problems can be found in the design of equipment mechanisms, parts materials and shapes, the construction and shapes of jigs, and detection systems (both the sensors and the systems themselves). Frequently, idling and minor stoppages are caused by mismatching jigs to workpiece shapes.

Another common cause is using existing equipment containing hidden defects without treating it as such. This produces frequent stoppages and makes adjustment procedures extremely difficult, with the frequency of stoppages depending on the skill with which adjustments are carried out. In such cases, look long and hard at the inherent weaknesses of the equipment and take steps to improve it. The problems making this kind of investigation necessary are not normally common to the whole range of products. They usually apply only to particular products.

The role of equipment design weaknesses in idling and stoppages should be explored last — after correcting slight defects and usage problems, applying P-M analysis to particular problems, and correcting unsatisfactory conditions. To do otherwise is to court failure. It is virtually impossible to pinpoint design weaknesses when they are mixed up with a variety of other factors that can be corrected through restoration. It is essential to follow the improvement steps in the proper sequence.

Key Points and Cautions

The following points should be kept in mind when conducting activities aimed at eliminating idling and minor stoppages:

Two Approaches Are Needed

Two approaches can be adopted to reduce idling and minor stoppages. The first approach is positive and has already been outlined: track down *all* the main causes and take steps to prevent stoppages from occurring. Reducing stoppages to zero is difficult, however, even when a variety of remedial measures are employed. The second approach is negative; it involves the creative use of detection techniques to detect and signal the occurrence of stoppages so they can be dealt with immediately.

Both approaches should be used together. In areas where operators manage many machines, stoppages are particularly difficult to detect without special warning systems.

Address Common Problems Before Particular Problems

Some problems are common to every machine regardless of the product or workpiece processed. Other problems are related to particular products or workpieces. In any program for reducing idling and minor stoppages, the best and quickest results are obtained by tackling common problems first. Particular problems require improvements to the equipment, such as better machines, jigs, and tools, and are likely to take time.

Beware of Complex Solutions

Problems often emerge at the interface between specific technologies and the systems in operation on the shop floor. Engineers tend to make judgments from limited technical standpoints, failing to consider in their analysis what actually happens on the shop floor. If this tendency is not avoided, solutions quickly become too complex and waste valuable improvement efforts.

Take Action Against Every Type of Occurrence

Idling and minor stoppages characteristically move from one location to another on a machine. Do not neglect any type of stoppage, no matter how infrequent its occurrence. In many cases, when attention is drawn exclusively to that portion of the machine where frequent stoppages occur, action is taken and the number of occurrences there drops. Stoppages then increase in another part, however, and there is no overall improvement. To avoid this, take action against stoppages wherever they occur in the machine, regardless of the relative frequency of occurrence.

Steps in Reducing Idling and Minor Stoppages

An overview of the improvement program for reducing idling and minor stoppages appears in Figure 3-15.

Case Study 3-4 — Reducing Blockages by Correcting Slight Defects

Idling and minor stoppages were occurring in a rolling machine using thread-rolling dies to make screws. Blockages were particularly common

- at the transfer point between the hopper and the chute
- on the chute
- at the transfer point between the chute and dies

Blockages occur as a result of high frictional resistance. To overcome this problem the improvement team redesigned the configuration of the transfer points by going back to first principles and eliminating all slight defects. The following improvements were carried out:

- The chute was cleaned (eliminating all dirt).
- The chute surface was improved (treating scratches, irregularities, wear, roughness).
- Chute installation was improved (treating gaps, clearances).
- The surfaces of butted plates were improved (uneven wear, irregularities).
- The attachment of butted plates was improved (controlling play, vibration, movement).

As a result of the team's efforts, idling and minor stoppages were reduced and higher operating speeds became possible (Figure 3-16; *see* p. 148).

Case Study 3-5 — Correcting Blockages

Idling and minor stoppages were occurring in machinery that supplies parts from a feeder, selects them on a chute accord-

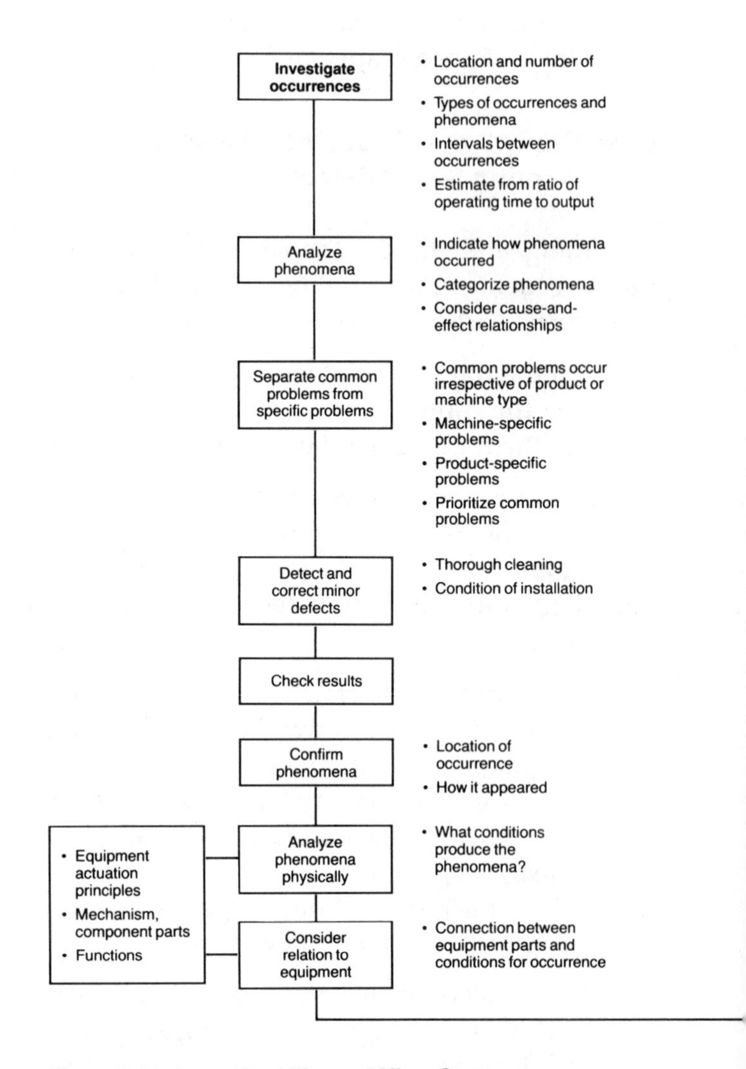

Figure 3-15. Improving Idling and Minor Stoppages

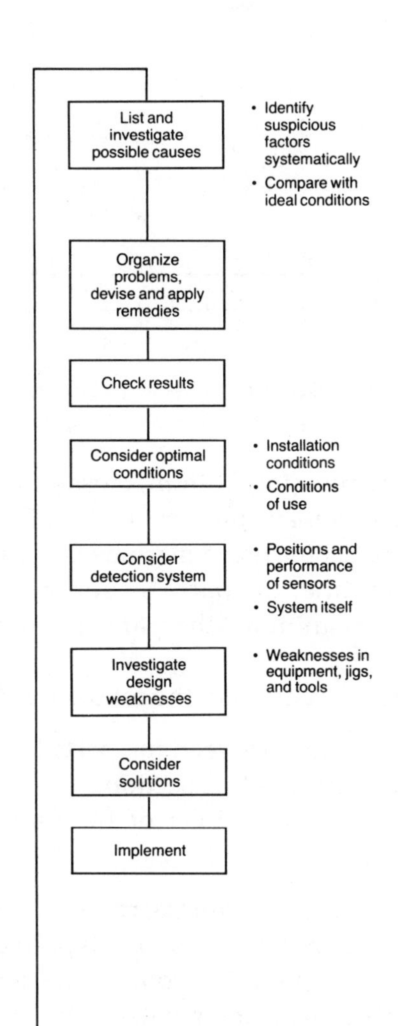

List and investigate possible causes
• Identify suspicious factors systematically
• Compare with ideal conditions
Organize problems, devise and apply remedies
Check results
Consider optimal conditions
• Installation conditions
• Conditions of use
Consider detection system
• Positions and performance of sensors
• System itself
Investigate design weaknesses
• Weaknesses in equipment, jigs, and tools
Consider solutions
Implement

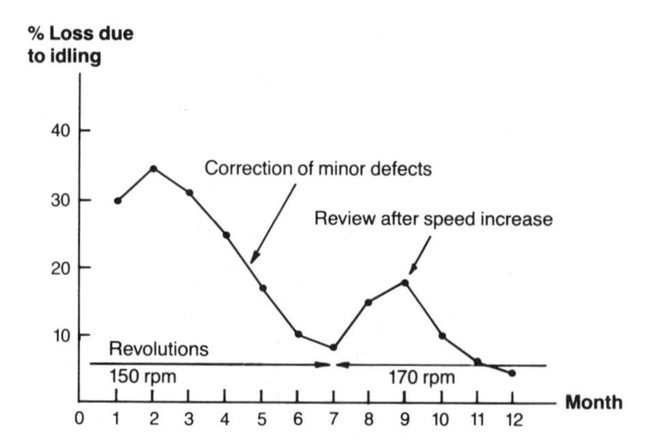

Figure 3-16. Reducing Blockages by Correcting Minor Defects

ing to their orientation (top/bottom, crosswise/lengthwise), and feeds them to an automatic processing machine. The cutting capacity of an automatic processing machine is one every 3.5 seconds, or 17 to 18 per minute; the parts feeder must supply at a rate of $18 \times (4 + \alpha)$ per minute (the parts must be sorted in four directions).

The stoppages occurred in the following points:

- the transfer point between the parts feeder and chute
- the orientation selecting section
- on the chute and at inlet of the automatic processing machine

The stoppages resulted from parts jamming because of their positioning at the transfer points, overlapping, and so on. This happened at high output rates from the feeder, when the parts emerged in a mass instead of separately. The following steps were taken to give a constant feed rate:

- A constant quantity of parts was kept in the feeder via an automatic supply system with buckets.
- The vibration amplitude of the chute was set to the optimal value to give a smooth flow of parts (at the required rate but without overlapping).
- Resonance was corrected.

Blockages in the orientation selecting section of the chute also occurred at high feed rates. They were greatly reduced by stabilizing the feed rate, as described above, and by improving the angles, positions, and shapes of the chute installation (Figure 3-17).

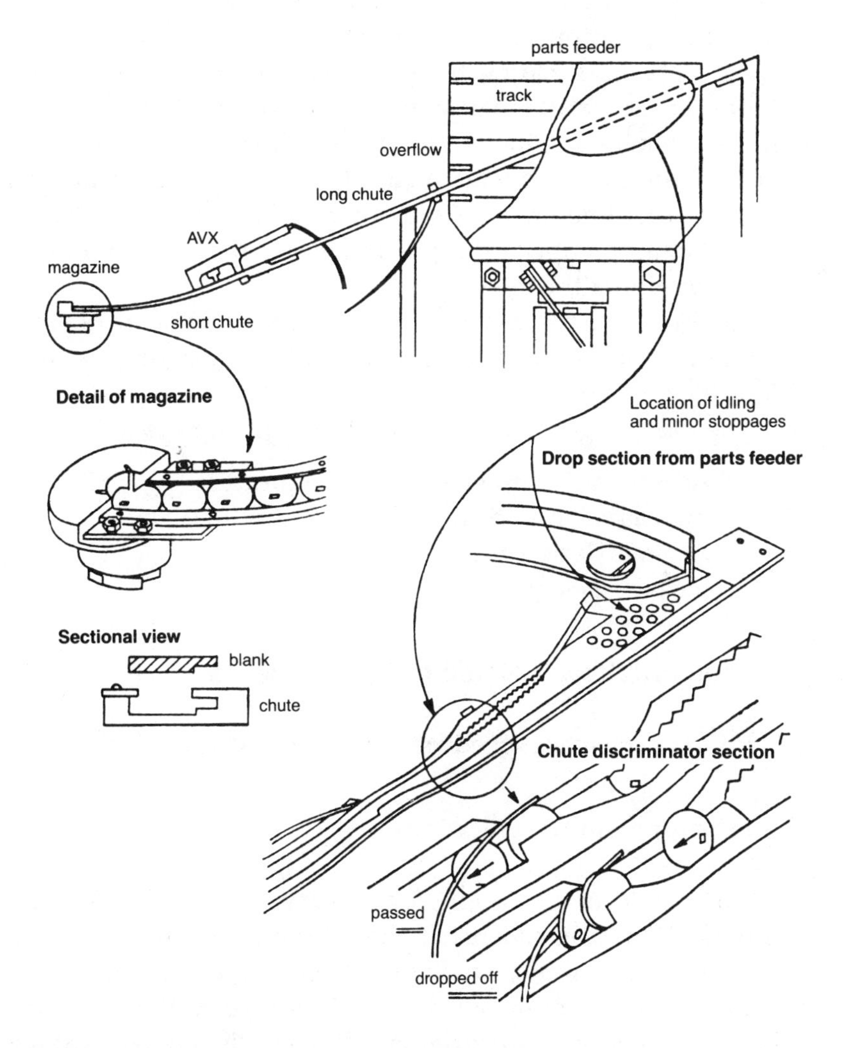

Figure 3-17. Improvements to Prevent Blockages

REDUCING SPEED LOSSES

A *speed loss* is the lost production caused by the difference between the designed (or standard) speed of a machine and its actual operation speed. This loss can be prevented by keeping the machine operating at the speed set by the operating standards. While this often proves impossible in practice, reducing speed losses can significantly increase the overall efficiency of the plant.

As a practical matter, there are often problems associated with the speeds set at the design stage. For example, lack of care may have generated inherent weaknesses in the design that prevent the equipment from maintaining the rated speed. Changes in the product line or increased complexity in product shapes since the equipment was designed may also prevent equipment from maintaining the rated speed.

For these reasons, a *standard speed* is set for each type of product. The standard speed is used instead of the designed speed to determine the speed loss.

The following steps are involved in reducing speed losses:

1. Achieve standard speed for each product.
2. Increase standard speed for each product.
3. Achieve design speed.
4. Surpass design speed.

Common Problems Related to Speed Losses

Efforts to increase speed are hindered by various problems:

Vague Equipment Specifications

Lack of care at the design stage may result in a speed specification that is unclear. As a result, the equipment is either operated beyond its speed limits, producing defects and breakdowns, or at a speed that is unnecessarily low. This is often the case with older equipment or in-house designs. It is usually profitable to take a fresh look at the speed specifications of these types of equipment.

Specified Speeds Are Attainable But Not Achieved

Some equipment may not be operated at the specified speed because of past quality or mechanical problems that were never solved. Such problems are often regarded as intractable; no effort is made to trace their real causes, which are often accelerated deterioration or slight defects left unchecked. The speed loss may be eliminated simply by correcting these minor problems.

In other cases, technical levels and control techniques have advanced since the equipment first experienced speed problems. Now, although it may cause no problems at current levels of in-house technology, the specified speed is still not being used.

Inadequate Investigation of Problems Exposed Through Speed Increases

As speeds are gradually increased over the present levels, quality or mechanical problems may appear immediately, or they may break out suddenly when a certain speed is reached. The defects causing these problems are latent at the lower speeds, surfacing only with an increase in speed.

Increasing the speed of a piece of equipment brings its latent defects to the surface all at once. Increasing speed is therefore a simple, productive way of exposing defects. Most companies, however, do not investigate and sort out the phenomena and problems produced by speed increases. They merely observe that defect rates, breakdowns, and adjustment frequencies are increased. The reasons for these increases are not tracked down, the problems remain hidden, and the operating speed is reduced to its former, less than optimal value.

Approaches to Increasing Speed

A vital first step, then, is to expose the hidden problems and determine whether they correspond to any of the following:

- unresolved defects due to insufficient debugging during the engineering stage

- defects in equipment mechanisms or systems
- inadequate daily maintenance
- insufficient precision, and so on

Once the causes are identified, measures can be devised to correct them. Solutions to these problems may help increase current technical capabilities and may have a beneficial effect on maintainability improvement and maintenance-prevention design.

Generally, improvement activities to increase speed should be organized with the same understanding, using the same methodology recommended for reducing breakdowns, idling and minor stoppages, and defects. The sections on these subjects should be consulted. A systematic improvement program for increasing speeds is outlined in Table 3-12.

Determine present levels	• Speed • Bottleneck processes • Downtime/frequency of stoppages • Conditions producing defects
Check difference between specification and present situation	• What are the specifications? • Difference between standard speed and present speed • Difference in speeds for different products
Investigate past problems	• Has the speed ever been increased? • Types of problems • Measures taken to deal with past problems • Trends in defect ratios • Trends in speeds over time • Differences in similar equipment
Investigate processing theories and principles	• Problems related to processing theories and principles • Machining conditions • Processing conditions • Theoretical values
Investigate mechanisms	• Mechanisms • Rated output and load ratio • Investigate stress • Revolving parts • Investigate specification of each part
Investigate present situation	• Processing time per operation (cycle diagram) • Loss times (idling times) • *Cp* value of quality characteristics • Check precision of each part • Check using five senses

Table 3-12. Strategies for Increasing Speed

List problems	• List problems and identify conditions that should exist • Compare with optimal conditions • Problems with mechanism • Problems with precision • Problems with processing theories and principles
List predictable problems	• Mechanical • Quality
Take remedial action against predictable problems	• Compare predictable problems with present conditions • Take action against predictable problems
Correct problems	
Perform test runs	
Confirm phenomena	• Mechanical • Quality • Change in *Cp* values
Review analysis of phenomena and cause-and-effect relationships and carry out remedial actions	• Physical analysis of phenomena • Conditions producing phenomena • Related causes
Perform test runs	

Table 3-12. Strategies for Increasing Speed

REDUCING CHRONIC QUALITY DEFECTS

When a production system regularly produces totally or partially defective products despite various improvement and control measures, these defective parts are termed *chronic* quality defects.

Irreparably defective products are obvious losses; less obvious are the losses generated by partially defective products requiring an additional investment of manhours in rework or repair. Because they can be repaired, partial defectives are often not counted as defects. In a successful program to reduce chronic losses, however, all defective outcomes should be met with equal concern.

General Characteristics of Chronic Quality Defects

To successfully reduce chronic defects, improvement teams must learn to recognize them and to avoid the most common traps.

Improvement Efforts Have Been Unsuccessful

Even the most determined efforts can seldom trace the causes of chronic quality defects. In desperation, quality teams may adopt trial-and-error measures without knowing the causes, but these often have no effect. Eventually, team members simply give up, and the problems remain uncorrected.

Various justifications are advanced for abandoning the effort to correct chronic quality defects: "Implementing effective preventive measures is impossible when we don't know the causes of the defects," or, "There's no way to solve this problem — it's inherent in the nature of the equipment." Usually, however, avoidable mistakes along the way have led the investigation astray.

The Problem Is Approached in the Wrong Way

Because we typically approach a problem with the goal of identifying its cause, we may jump to the wrong conclusions or narrow down the causes too quickly. Then we develop remedies to address the few causes we have identified. Unfortunately, chronic quality defects are often produced by an ever-changing combination of causes. Every suspicious factor must be tackled, because little progress is possible when only a few are pursued.

Thinking Is Limited to Specific Technical Fields

Top staff engineers in most companies are experts in particular technical fields. In problem-solving for chronic defects, they tend to overlook solutions outside their own areas of expertise and favor complex over simple solutions. As a result, many problems remain unsolved.

For example, consider the problem of broken punches in a punch press machine. An engineer might propose changing the heat-treatment conditions or increasing the average hardness of the punches without considering the wide range of punch lifetimes — why it exists, where it originates, or how it might be addressed. People are often attracted to the more interesting technical solutions, overlooking the reality that the causes of dispersion lie in much simpler factors — the way the punches are installed, the installation precision, the machine precision itself, and surface roughness of the punches.

The solutions to some chronic quality defects can be found in specific technical areas, but most problems require a wider viewpoint that considers causes in actual operations on the shop floor.

In practice, a single defect phenomenon often has many interdependent causes, all of which are likely to undergo changes. The phenomenon can occur immediately as a result of any of these changes. For example, setup produces different results depending on the method used, the way adjustments are carried out, gap dimensions, installation methods, the setting of processing conditions, and so on. Whether defective products are produced depends on the combination of factors in the setup.

Engineers must learn to spot the variables related to operations and equipment by observing and studying

- actual operations carried out on the shop floor
- setup and adjustment operations
- the equipment itself

When determined efforts to find technical solutions do not produce the expected results, or when the results vary widely, the actual causes of quality defects on the shop floor are probably not constant. Solutions that take into account the interface between engineering and the shop floor are more likely to succeed.

Identifying and Investigating Causes Is Difficult

Quality improvement teams encounter two common problems: mistaken identification of the causes of chronic quality defects and inadequate investigation of causes once they are correctly identified.

First, true causes are obscured by deficiencies in the observation and analysis of defect phenomena, mistakes in analytical methods, thinking based solely on past experience, and jumping to hasty conclusions.

Second, improvement efforts are thwarted by lack of thoroughness and other deficiencies in the way managers and engineers observe equipment and work methods. Hidden defects are not recognized as such and causes that are noted are disregarded as having no effect on the phenomena being investigated.

As a result of these deficiencies, hidden defects and their warning signs often go unrecognized. When these causes are left unchecked, quality defects recur. Real progress in reducing chronic quality defects requires an entirely new approach to the discovery and investigation of hidden defects. Figure 3-18 shows the relationships among the typical problems associated with chronic quality defects.

Sporadic Problems Versus Chronic Problems

According to J. M. Juran, sporadic defects result from sudden, adverse changes in the status quo, for example, in current control points or causal factors. They require treatment that restores the status quo, bringing the variables back to their original state — by replacing a worn tool, for example (Table 3-13).

Chronic defects result from adverse conditions that have been accepted as normal over time. They require breakthrough solutions that change the status quo. Breakthrough thinking is needed to tighten the control ranges of existing factors, develop control methods that prevent even slight defects from escaping notice, and give special consideration to factors not presently controlled.

Strategies for Reducing Chronic Problems

Solving sporadic problems requires an essentially conservative approach involving the following strategies:

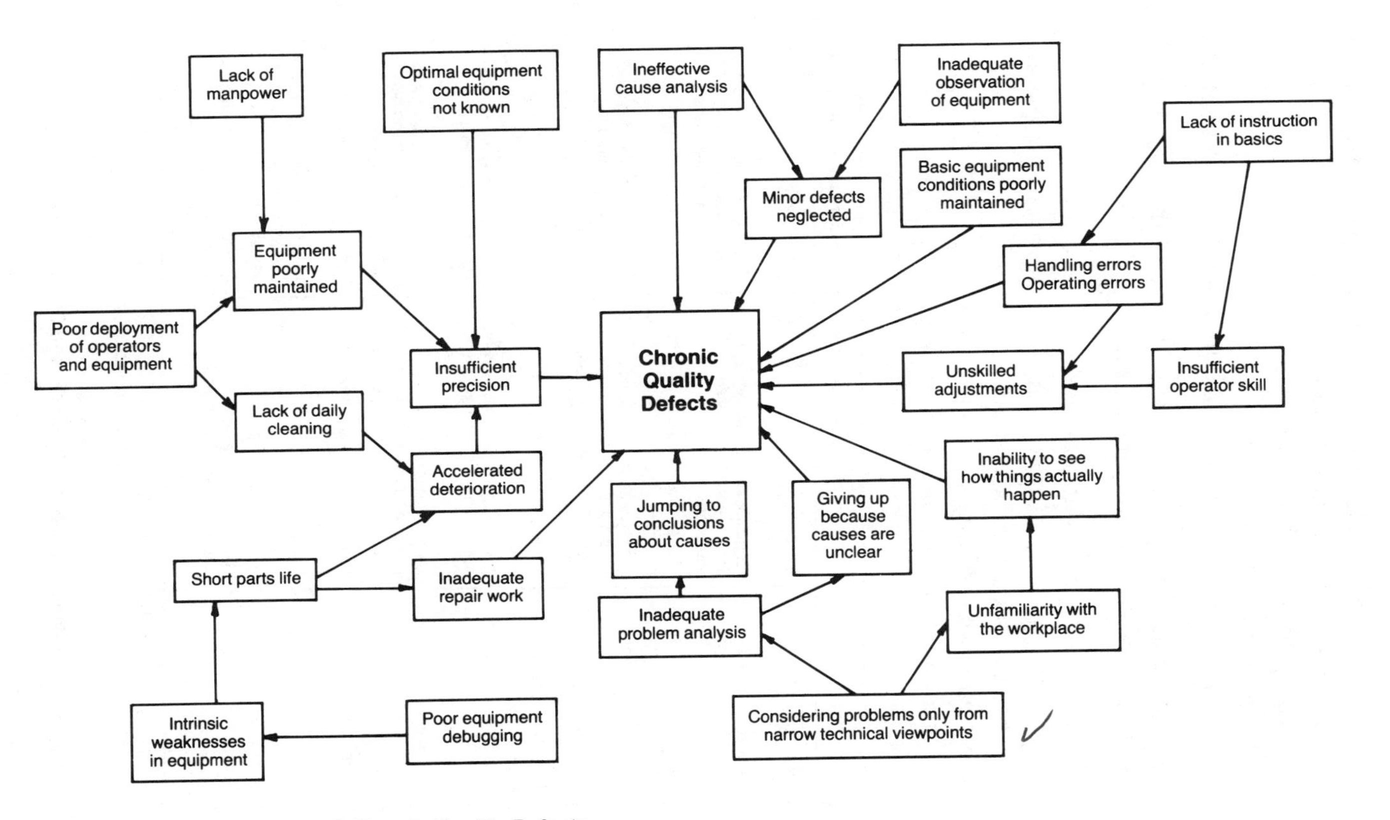

Figure 3-18. Typical Causes of Chronic Quality Defects

- restoration by maintaining or controlling current conditions
- targets set to current standards
- comparisons with current standards
- checking control points
- responsibility on the part of both operators and managers

Solving chronic problems requires breakthrough thinking involving a different set of strategies:

- improvements that change the status quo
- targets set according to company improvement goals
- review of existing standards
- review of existing control points
- responsibility on part of managers

Aspect	Nature of sporadic defects	Nature of chronic defects
Tangible economic loss	Minor	Major
Extent of irritations caused	Substantial. Sudden nature of trouble attracts supervisory attention	Small. Continuing nature of trouble leads all concerned to accept it as unavoidable
Type of solution required	Restore the status quo	Change the status quo
Types of data needed	Simple data showing trend of quality with respect to one or two variables such as time or lot number	Complex data showing relation of quality to numerous variables
Plan for collecting data	Routine	Specially designed
Data collected...	By inspectors, service representatives, etc., in the usual course	Often through special experimental procedures of their work
Frequency of analysis	Very frequent. May require review every hour or every lot	Infrequent. Data may be accumulated for several months before analysis is made
Analysis made by...	Line people such as a design or production supervisor	Technical personnel
Type of analysis...	Usually simple	Possibly intricate. May require correlation study, analysis of variance, etc.
Action by whom...	Usually by line personnel in design, manufacturing, etc.	Usually by personnel other than those responsible for meeting the standard

From *Quality Planning and Analysis: From Product Development through Usage* by J.M. Juran and F.M. Gryna, Jr. (New York: McGraw-Hill Book Company, 1970), 9

Table 3-13. Distinguishing Between Sporadic and Chronic Defects

Stabilize Causal Factors

To reduce chronic problems we must stabilize all variable factors, identify significant differences between normal and abnormal conditions, and study ways to prevent defects from being generated in the first place.

Causal factors are all factors that might conceivably affect results (*i.e.*, the defect phenomena), including those that are logically proved to produce the phenomena. *Causes* are those causal factors that are proven or deduced to produce the phenomena, directly or indirectly.

To *stabilize* something is to prevent it from changing; stabilizing causal factors means preventing them from changing.

Although causal factors may appear stable in factories and workshops, work is actually carried out under extremely unstable conditions, in a tangled mass of varying causal factors. Remedial action taken while causal factors are changing is likely to fail. Moreover, the actual results of such action are impossible to judge.

To reduce chronic problems, these tangled, fluctuating variables must be stabilized one by one. For example, a single operation may proceed differently from day to day, depending on the operators who carry it out. Their adjustment methods and judgments, the ranges they use, and the mistakes they make when dealing with quality abnormalities will vary considerably.

This variability is caused by lack of standardization in the workplace or failure to observe set standards. In such circumstances, managers may be unaware or negligent; operators may not appreciate the significance of certain aspects of their work, or they may assume that their methods are correct.

The only realistic solution is to pinpoint the sources of trouble through a process of elimination by stabilizing the causal factors one by one. Stabilize every causal factor that might logically have an effect on the outcome:

- processing principles
- mechanisms
- operation and adjustment
- precision of equipment, jigs, and tools
- work methods

Comparative Studies

In any program to reduce quality defects, normal conditions (no defects) should be compared systematically with abnormal conditions (defects) to identify significant differences. Problems that would be easy to solve with simple solutions very often remain unsolved when this step is omitted, even if the defects and phenomena are acknowledged.

Once comparative studies identify the location, nature, extent, and causes of significant differences, they should be analyzed quantitatively and qualitatively. Several methods can be used:

Compare products (results). Compare defective and non-defective products in terms of shape, dimensions, and functions. Also investigate the variation in defects over time and in terms of their location in the product.

Compare processes. Compare the machines, jigs, tools, and dies producing defective products with equipment producing good products to identify any differences in shape, dimensions, surface roughness, and so on. Make a special effort to develop new measuring methods for factors that do not appear to be quantifiable.

Compare the effects of changing parts. With assembled products, study the effects of interchanging parts that might be related to the defects. Also exchange machines parts, jigs, and tools to determine any differences.

Comparative studies should be used to organize the following factors:

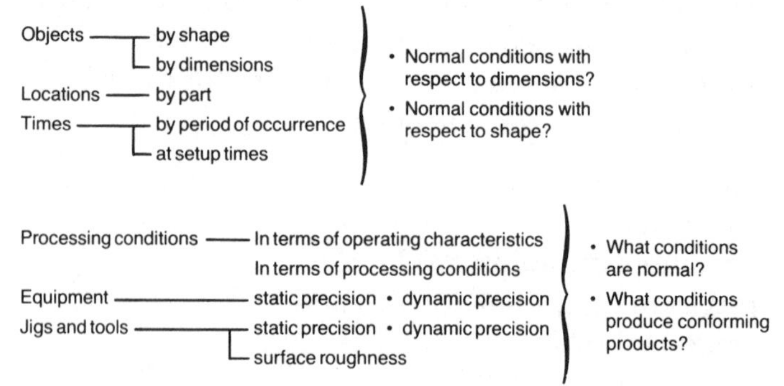

To increase the effectiveness of comparative studies consider the following methods:

Increase analytical precision to detect more subtle differences. Increase analytical precision by using a magnifying glass, microscope, or other apparatus to detect differences too small to be seen with the naked eye.

Nonquantifiable factors can pose obstacles; consider analyzing nonquantifiable shape differences.

Investigate new measuring methods. Often measurements are not performed because appropriate methods have not been developed. If that is the case, consider developing new measuring methods. New measurements can often clarify significant differences. Even if the differences are not made completely clear, telltale signals showing their existence can often be detected.

For example, surface roughness gauges and microscopes with projectors can be effective when localized uneven wear and surface roughness do not show up as dimensional differences.

Review Causal Factors

Review the current causal factors and control points and consider a new approach to their selection and study. The best approach is P-M analysis: conduct a physical analysis of the phenomena, identify the conditions needed for the occurrence of the phenomena, list the relevant causal factors and sources of trouble, and analyze them (*see* Chapter 2).

Study the Relationship Between Equipment Parts and Quality Characteristics

Analyzing the phenomena actually creating defects is important in defect reduction, but it cannot promise continuing improvement. Because these phenomena suddenly appear and disappear, or are replaced by different phenomena, we should also consider methods that prevent defects from occurring in the first place. This is even more advisable for products that, while not actually defective, barely meet standards and have large differences in quality or low C_p values (process capacity indices).

To reduce defects, a deductive as well as an analytical approach to investigation should be adopted. The analytical approach, already presented, focuses on the status quo, while the deductive method approaches from first principles the functions of equipment components, jigs, and tools. These are considered in relation to the quality characteristics of the product, with the aim of determining what their basic functions should be and how they should be configured. For example:

How do the precision and shapes of parts, jigs, and tools affect and relate to the quality characteristics? Are the qualitative and quantitative relationships between equipment parts, jigs, and quality characteristics fully understood? How close are these relationships? What are the current limits on the precision required to maintain the quality characteristic values?

What effect do the precision and shapes of parts, jigs, and tools have on the C_p values, separately and in combination? The precision and shapes of parts, jigs, and tools may affect the C_p values, both separately and in combination. Have these relationships been considered fully? Are the factors controlling the C_p values and their quantitative relationships fully understood?

What is the optimal configuration considering the functions of parts, jigs, and tools? Has the configuration required to satisfy the quality characteristics been clarified? Have the required static and dynamic precision and the external shapes of the parts been considered? What would be the effect on the quality characteristics if these were to change?

These issues must be investigated to determine the current state of equipment functions and configurations and whether they are being correctly maintained. Through a comparison of the present state of the equipment with the ideal state of the equipment, continual improvement targets can be identified (Table 3-14 and Figures 3-19 and 3-20).

REFERENCES

Shingo, Shigeo. *A Revolution in Manufacturing: The SMED System.* Cambridge: Productivity Press, 1985.

Shirose, Kunio. "Minor Stoppages" (in Japanese). *Plant Engineer* 13 (April and May 1981): 44.

Main Components	Functions	Targeted Function	Relation to Quality Characteristics					Relation To Optimal Conditions	Checking Methods
			Q_1	Q_2	Q_3	Q_4			

Table 3-14. Relation between Main Parts and Quality Characteristics

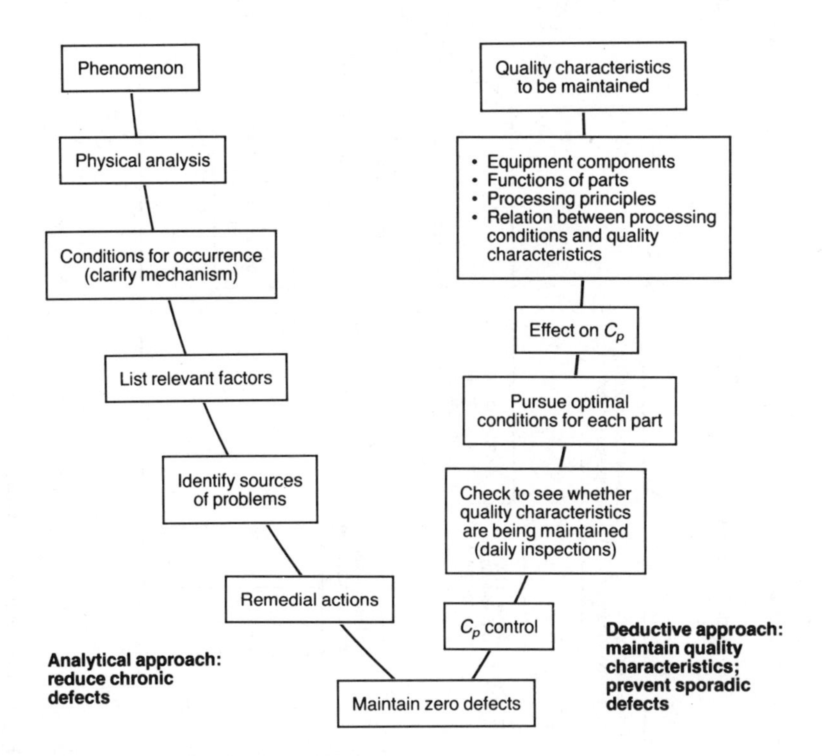

Figure 3-19. Analytical and Deductive Approaches

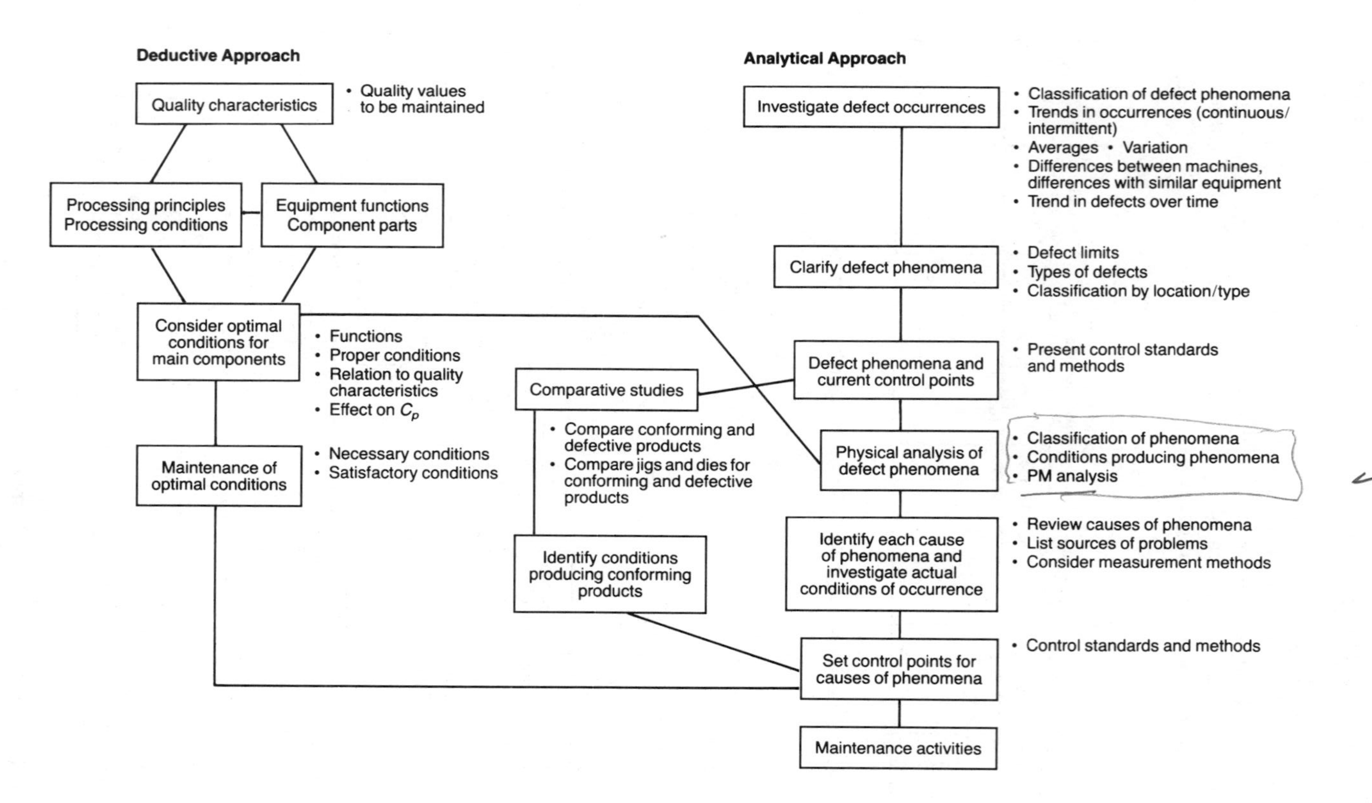

Figure 3-20. Improving Quality

4
Autonomous Maintenance

Ideally, whoever operates equipment should maintain it, and originally, these two functions were combined. Gradually, however, the maintenance and production functions were separated as equipment became more sophisticated, as businesses grew larger, and as American-style PM was widely adopted. During Japan's postwar period of rapid industrial growth, most equipment was replaced by newer, unfamiliar equipment. Responding to demands for increased production, production departments concentrated on output, while maintenance departments gradually assumed responsibility for almost all maintenance functions. The resulting bipolar specialization continues today.

During the present low-growth era, however, companies are increasingly under pressure to boost competitiveness and reduce costs. Today, many managers are keenly aware that a decisive factor in increasing competitiveness is more efficient equipment utilization.

Maintenance performed by equipment operators, or *autonomous maintenance*, can contribute significantly to equipment effectiveness. At the heart of autonomous maintenance is *deterioration prevention*, which has been neglected in most factories until recently. Considering the importance of this activity in manufacturing today, it is surprising that autonomous maintenance was not promoted earlier.

OPERATION AND MAINTENANCE ARE INSEPARABLE

Efficient production depends on both production and maintenance activities, but the relationship between operators and maintenance personnel is often somewhat adversarial. No matter how hard maintenance personnel work, they can make little progress in maintenance and equipment improvement as long as the operator's attitude toward maintenance is "I operate — you fix."

If, on the other hand, operators can participate in the maintenance function by becoming responsible for the prevention of deterioration, maintenance targets are more likely to be achieved. This cooperative effort allows maintenance personnel to focus their energies on tasks requiring their own technical expertise; it represents the first step toward more efficient maintenance.

The two departments must do more than share the responsibility for equipment — they must work together in the spirit of cooperation. Maintenance cannot simply wait passively for orders from the production department. Nor can production expect miracles, when maintenance is overwhelmed with work orders. Operators are responsible for production, and it is only human for them to become impatient if repairs are not done right away. There is no way maintenance targets will be achieved, however, if the two groups fail to understand each other's situation or, in extreme cases, if they are at odds with one another.

CLASSIFICATION AND ALLOCATION OF MAINTENANCE TASKS

This section classifies maintenance activities and allocates tasks in the autonomous maintenance program (Figure 4-1; *see* pp. 168-169).

Two types of activities are required to increase equipment effectiveness:

1. *Maintenance activities* prevent breakdowns and repair ailing equipment. They occur in a cycle consisting of normal operation combined with preventive maintenance (*i.e.*, daily, periodic, and predictive maintenance) and corrective maintenance.
2. *Improvement activities* extend equipment life, reduce the time required to perform maintenance, and make maintenance unnecessary. Reliability and maintainability improvement, maintenance prevention, and maintenance-free design are all maintenance improvement activities.

Maintenance and improvement activities must be carried out simultaneously in three areas of deterioration: prevention, measurement, and restoration. Maintenance goals cannot be achieved if any one of these areas is neglected; the methods used and the priority, however, may differ from department to department or factory to factory.

Although deterioration prevention is the most basic maintenance activity, it is often neglected in favor of periodic inspection and precision tests. If efficient maintenance is the goal, however, this practice is rather like putting the cart before the horse.

Program for the Production Department

The production department must carry out the following three deterioration-prevention activities:

1. *Deterioration prevention*:
   - Operate equipment correctly.
   - Maintain basic equipment conditions (cleaning, lubrication, bolting).
   - Make adequate adjustments (mainly during operation and setup).
   - Record data on breakdowns and other malfunctions.
   - Collaborate with maintenance department to study and implement improvements.

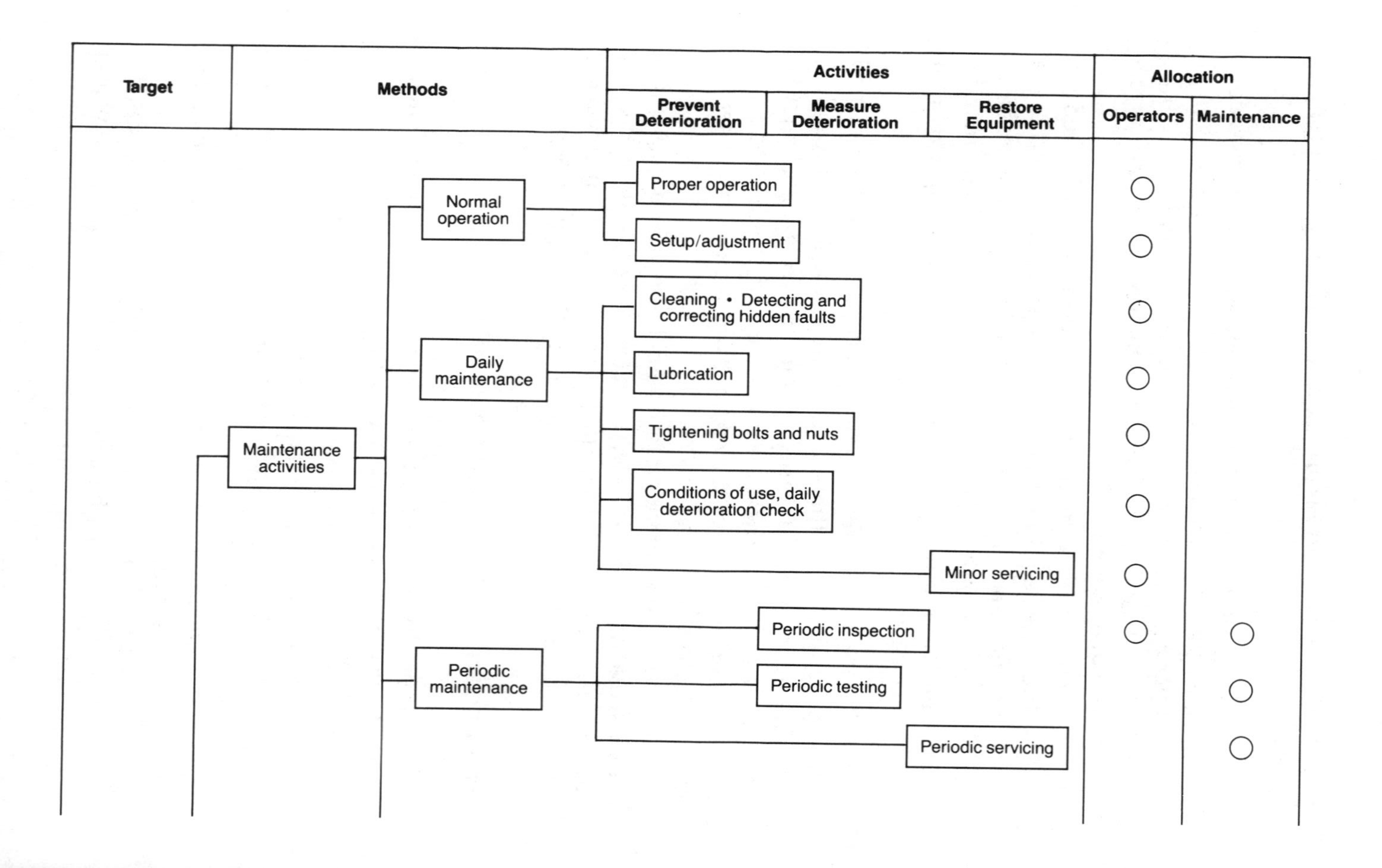

Target
Methods
Activities
Prevent Deterioration
Measure Deterioration
Restore Equipment
Allocation
Operators
Maintenance
Maintenance activities
Normal operation
Daily maintenance
Periodic maintenance
Proper operation
Setup/adjustment
Cleaning • Detecting and correcting hidden faults
Lubrication
Tightening bolts and nuts
Conditions of use, daily deterioration check
Minor servicing
Periodic inspection
Periodic testing
Periodic servicing

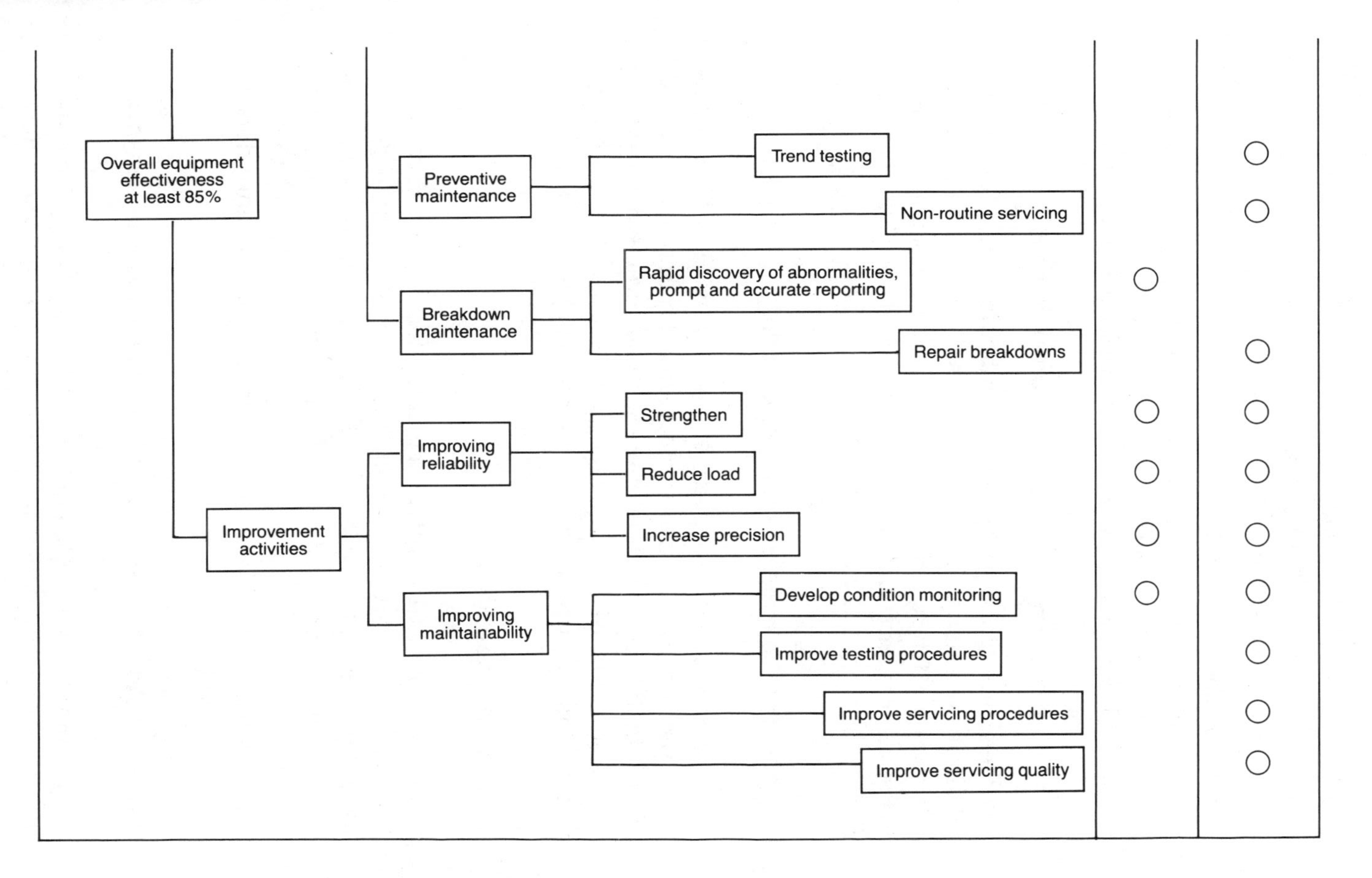

Figure 4-1. Classification and Allocation of Maintenance Tasks

2. *Deterioration measurement* (using the five senses):
   - Conduct daily inspections.
   - Conduct certain periodic inspections.

3. *Equipment restoration*:
   - Make minor repairs (simple parts replacement and temporary repairs).
   - Report promptly and accurately on breakdowns and other malfunctions.
   - Assist in repairing sporadic breakdowns.

These activities, particularly maintaining basic equipment conditions (cleaning, lubrication, bolting) and daily inspection, help prevent deterioration but cannot be addressed adequately by the maintenance staff alone. They are most effectively handled by those closest to the equipment — the operators.

Program for the Maintenance Department

The maintenance department performs periodic maintenance, predictive maintenance, maintainability improvement, and other activities involving deterioration measurement and equipment restoration. Maintenance specialists should be concentrating their efforts on work requiring a high level of technical skill.

Improve Maintainability

An important but often neglected focus for maintenance departments is improving maintainability. Production departments typically fight to cut seconds from working times; by comparison, the attitude and activities of maintenance departments toward improving maintenance efficiency are poor. Correcting this situation should be a top priority for maintenance department managers.

Guide and Assist Operators with Autonomous Maintenance

Autonomous maintenance can only become established with the proper guidance and assistance from the maintenance department. Maintenance departments often overlook the need to instruct operators in the maintenance procedures they are asked to carry out. For example, they may request daily inspections and prepare and enforce inspection standards without teaching inspection methods, set standard lubrication procedures that take 30 minutes when only 10 minutes are available at startup, or require inspection and lubrication without helping to develop easier ways of doing them.

When autonomous maintenance is not progressing, maintenance personnel should consider whether they have given adequate guidance and instruction in turning over maintenance responsibilities to operators.

Other Activities

Other important activities for the maintenance department include

- research and development of maintenance technology
- setting maintenance standards
- keeping maintenance records
- evaluating results of maintenance work
- cooperating with engineering and equipment design departments

ESTABLISHING BASIC EQUIPMENT CONDITIONS

Establishing basic equipment conditions is an important activity in autonomous maintenance. This activity includes cleaning, lubrication, and bolting.

Cleaning and Cleanliness

As the word implies, cleaning means removing dirt, stains, dust, chips, and other foreign matter adhering to machines, dies, jigs, tools, raw material, workpieces, and so on. During this activity operators also look for hidden defects in their equipment and take action to treat them.

Harmful Effects of Inadequate Cleaning

The harmful effects of inadequate cleaning are too numerous to list here. Typically, however, they appear directly or indirectly in the following ways:

- Foreign particles enter sliding machine parts, hydraulic systems, or electrical systems, producing frictional resistance, wear, clogging, leakages, and electrical faults. This causes losses in precision, equipment malfunctions, and breakdowns.
- In certain types of automatic equipment, the presence of particles or dirt on the supply chutes or workpieces adversely affects the flow of work, causing malfunctions, idling, or minor stoppages.
- Often product quality is affected directly. For example, in plastic extrusion molding machines, foreign matter adhering to the dies or other attachments or contaminating the plastic pellets (feed material) causes carbonizing inside the cylinder or other attachments or causes resin to leak from the die-mounting face. This interferes with the proper resin flow, makes die changeovers difficult, or causes the resin to burn and stick.
- In the assembly of relays and other electrical control parts, dirt and dust on jigs and tools adhering to contacts causes lethal electrical faults.
- In precision machining, dirt adhering to jigs, tools, and their mountings slows down centering operations and causes eccentricity during machining, producing defective products.
- In electroplating, contaminated workpieces or dirt or foreign particles in the electrolyte can produce defects in the plating.

Defects in dirty equipment are hidden for both physical and psychological reasons. For example, wear, play, scratches, deformation, leaks, and other equipment defects may be concealed on dirty equipment. Moreover, operators may have some psychological resistance to inspecting dirty equipment carefully.

Cleaning Is Inspecting

Cleaning is not simply making the equipment look clean, although it has that effect. Cleaning also means touching and looking at every part of the equipment to detect hidden defects and abnormalities such as excess vibration, heat, and noise. In other words, *cleaning is inspecting*. In fact, if cleaning is not done in this way it loses its meaning.

When operators thoroughly clean a machine left to run by itself for a long time, they may find as many as 200 to 300 defects, occasionally including serious defects presaging serious breakdown.

Dirt, dust, play, abrasion, surface damage, looseness, deformation, and leaks in machinery, dies, jigs, and tools combine synergistically to cause deterioration and running problems. Cleaning is the most effective method of detecting such faults and preventing trouble. Checkpoints for cleaning are listed in Table 4-1.

Activities That Encourage Equipment Cleaning

Operators engage in three types of activity that promote cleaner equipment: they gain greater awareness and respect for their equipment by giving it an initial thorough cleaning; they eliminate the sources of dirt and contamination and make equipment easier to clean; and they develop their own cleaning and lubrication standards.

Start with Initial Cleaning

Cleaning equipment and touching every part of it can be a new experience for the operator. The activity yields many discoveries and questions. Although operators may do the work grudgingly at first, subsequent TPM group meetings and the cleaning itself will naturally encourage them to keep equipment clean, if only because they worked so hard to get it that way.

Many questions will surface:

- What kinds of malfunctions (quality or equipment) will occur if this part is dirty or dusty?
- What causes this contamination? How can it be prevented?
- Isn't there an easier way to clean this?
- Are there any loose bolts, worn parts, or other defects?
- How does this part function?
- If this part broke, would it take a long time to repair?

TPM operator groups follow up these questions as they arise and every member takes part in addressing them. This kind of group problem-solving helps foster the growth of autonomous maintenance.

1. Cleaning Main Body of Equipment	a. Check for dirt, dust, oily sludge, scraps, and other foreign matter adhering to equipment • Sliding parts, parts contacting workpiece, positioning parts, etc. • Frames, beds, conveyors, transfer lines, chutes, etc. • Gauges, jigs, dies, and other assembled parts of equipment b. Check for loose or missing nuts, bolts, etc. c. Check for play in sliding parts, jig fittings, etc.
2. Cleaning Ancillary Equipment	a. Check for dirt, dust, grease, scraps, and other foreign matter adhering to equipment • 10 Air cylinders, solenoid valves, 3-unit FRLs • Microswitches, limit switches, proximity switches, photoelectric tubes • Motors, belts, covers, and their surroundings • Surfaces of instruments, switches, control boxes, etc. b. Check for loose or missing nuts, bolts, etc. c. Check for buzzing in solenoid valves and motors
3. Lubrication	a. Check for dirt, dust, and sludge on lubricators, grease cups, lubricating devices, etc. b. Check lubricant levels and drip feed c. Cap all lubricating points d. Make sure lube pipes are clean and leak-free
4. Cleaning around equipment	a. Make sure tools are in their assigned places and that none are missing or damaged b. Check for bolts, nuts, etc., left on the machine c. Check all labels, nameplates, etc., for cleanliness and legibility d. Check all transparent covers, windows, and view plates for dirt, dust, and misting-up e. Make sure all piping is clean and leak-free f. Check surroundings for dirt and dust and for dust fallen from top of equipment g. Check for dropped parts, workpieces, etc. h. Check for defective workpieces left lying around i. Clearly separate conforming products, defective products, and scrap

Table 4-1. Cleaning Check Points

5. Treat causes of dirt, dust, oil leaks, etc.	a. Are the causes of dirt, dust, oil leaks, etc., clearly shown on a chart? b. Is action being taken to prevent the generation of dirt and dust? c. Is action being taken to prevent oil leaks and other types of leaks? d. Are there plans to deal with longstanding problems? e. Have any causes been overlooked?
6. Improving access to hard-to-reach areas	a. Are inaccessible areas clearly shown on a chart? b. Are there any special cleaning tools or other signs of ingenuity and effort? c. Have covers been made easier to remove to facilitate cleaning? d. Are there plans to deal with longstanding problems? e. Have any inaccessible areas been overlooked? f. Is everything kept tidy and in good order to facilitate cleaning?
7. Cleaning Standards	a. Are there separate standards for each piece of equipment or area? b. Have cleaning duties been clearly assigned? c. Are the types of cleaning and areas to be cleaned classified? d. Have cleaning methods and tools been specified? e. Have cleaning times and intervals been specified? f. Are the standards clear and easily understood by everyone? g. Are the cleaning times appropriate? h. Can the cleaning be completed within the times specified? i. Are all important cleaning items included? j. Is too much time specified for cleaning less important areas? k. Are inspection points that can be covered during cleaning clearly described?

Table 4-1. Cleaning Check Points

Base training on operators' questions. Authoritarian-style training divorced from the operators' activities has little effect. Discovery-based training is more effective. It answers questions that surface naturally when operators clean their equipment and links the answers to the next stage of action.

Emphasize important objectives of cleaning. In promoting equipment cleaning, supervisors and group leaders should emphasize the following points:

- the importance of basic equipment conditions (requirements) and how to achieve them (*e.g.*, cleaning, lubrication, and bolting)
- major cleaning checkpoints
- the meaning of the phrase "cleaning is inspection"

Prevent Contamination and Make Cleaning Easier to Perform

The more effort operators initially put into cleaning equipment, the more they want to maintain its hard-won cleanliness. At the same time, operators begin to feel a desire for improvement. For example, they may say:

- "No matter how many times I clean this part, it becomes dirty again quickly — what can I do to prevent that?" Or, "I can't bear to spend this much time on cleaning and lubricating — something must be done about it!"
- "The problems I managed to find and correct will crop up again if we just leave them like that. We ought to be doing regular inspections."
- "I've finished the cleaning for the moment, but I need help to cut down the number of breakdowns and faults."

Supervisors and group leaders can make good use of operators' ideas and eagerness to improve equipment. Actively encourage operators to propose practical ways of improving the equipment and to learn particular improvement methods. Their involvement will increase the satisfaction they feel when improvements are successful and build their confidence to go on to the next step.

Eliminate sources of contamination. After the initial thorough cleaning, sources of dirt, dust, and foreign matter, and their effects on equipment and product quality are easy to see. Their harmful effects have already been mentioned; eliminating their sources is a prerequisite for shortening cleaning times and preventing future problems.

There are various sources of contamination, such as chips, filings, dust, flashes, scale generated in processing, foreign particles contained in delivered materials, oil, water, and dust generated by equipment, or dirt and dust coming in from outside. Taking action against these contaminants means suppressing their sources, preventing dust and dirt from spreading, and preventing their infiltration into machinery with the use of covers and sealings, and so on.

Improve access to all areas to be cleaned. Inaccessible areas are time-consuming to clean or lubricate. If sources of contamination cannot be completely suppressed, improve cleaning methods so that cleaning takes as little time as possible.

Improve methods. Often after identifying all the points needing lubrication, operators discover that they don't have enough time to lubricate them all. They must find ways to overcome this obstacle. Many inspection standards prepared by the engineering staff disregard actual shop-floor conditions; operators must propose improvements based on the reality of the shop floor.

To deal with these and other related problems, promote the attitude that the shop floor is responsible for its own improvements. The engineering staff should then assist by answering questions raised at the shop-floor level. This approach to improvement generally turns out to be the quickest and most practical.

Think through and evaluate improvement plans. Sometimes costly plans are proposed that eliminate only a few minutes from cleaning time because of the difficulty of the cleaning operation. This approach to improvement is not cost-effective and ought to be discouraged.

Check results. Remember that even if their object is to reduce cleaning times, these types of improvements may also affect product quality, breakdown frequencies, setup, maintainability, and other factors. Study the ideas suggested by TPM autonomous maintenance groups from this perspective and expose their hidden, unexpectedly beneficial effects on these factors.

For example, a rubber-curing press (consisting of eight individual presses in a single frame) was wasting hydraulic fluid at rates of 300 to 500 liters per month, although there were no leaks from the piping, rams, valves, and so on.

Thorough initial cleaning revealed cracks in the fillet weld-zone at the bottom of the cylinders. Dismantling and repairing these cracks reduced the amount of hydraulic fluid consumed (Figure 4-2).

In another case, workpieces were falling off a production line (cure-molding, painting, and assembly), mainly while being conveyed between processes. Naturally, it took time to sweep

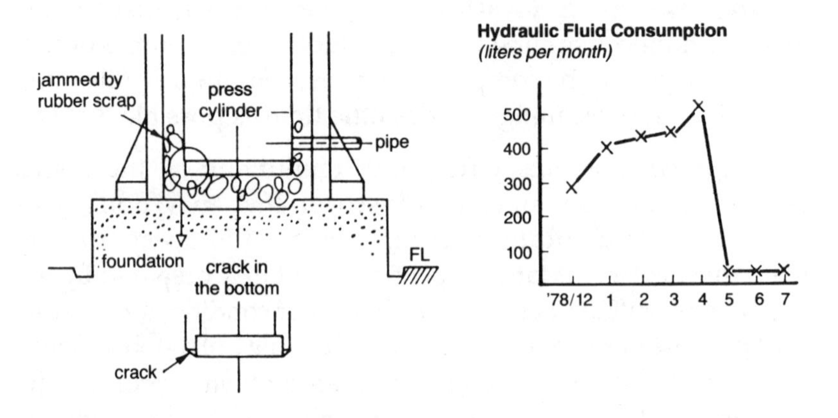

Figure 4-2. Eliminating the Source of Hydraulic Fluid Leaks in Line Press

the workpieces up, and they had to be discarded as defective. Along with actions against breakdowns, the autonomous maintenance group chose to look for ways of preventing this problem as the theme of their step 2 improvement activities. Their success is documented in Figure 4-3.

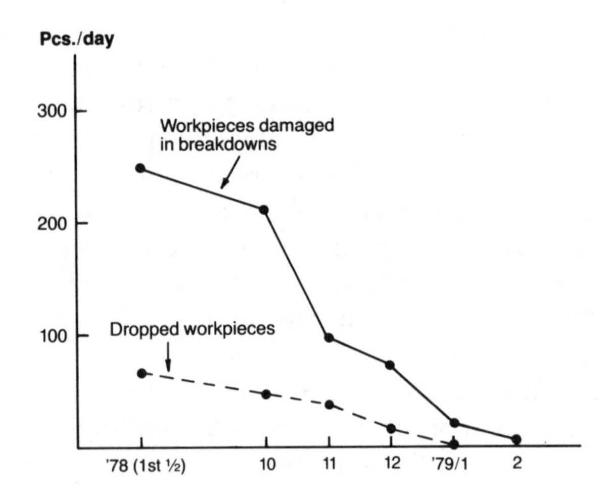

Figure 4-3. Preventing Dropped Workpieces

Prepare Cleaning and Lubrication Standards

Using the experience gained through cleaning and preventing contamination, operators identify the optimal conditions (cleaning, lubricating, bolting) for their equipment. The autonomous maintenance group must then set the operating standards required to maintain these conditions.

Shop-floor supervisors, attempting to promote the ideas of cleaning, lubrication, tidiness, and organization throughout the workplace, invariably come back half-despairing, with questions like this one: "I've tried any number of times to make operators follow the standards, but they won't. If there's a good method of making them do it, I want to know it."

Why operators don't follow standards. These supervisors have not really understood *why* operators fail to adhere to the standards. The biggest obstacle to adherence occurs when those who set standards are not the same people who must follow them. This promotes the attitude that "I (supervisors or engineering staff) set the standards and you (operators) obey them." When supervisors see standards as rules that must be obeyed, they typically overlook the need to explain why they are necessary, or how to follow them properly or to provide enough time. Rather than trying to force operators to follow standards, supervisors should support their effort in the following ways:

- Clarify the standards and how to follow them.
- Explain clearly *why* standards should be followed, that is, what will happen if they are not followed.
- Make sure operators have the necessary skills to follow the standards.
- Provide the necessary environment by making certain there is sufficient time, for example.

In other words, if the motivation, ability, and opportunity are not present, standards cannot be obeyed no matter how hard the supervisor tries to enforce them. Most activities related to autonomous maintenance depend on the skills and motivation of the operators actually performing them. Most supervisors will experience the anxiety of knowing what ought to be done but not communicating it adequately to the operators.

Self-set standards are best. The best way to ensure adherence to standards is to have them set by those who must follow them. Indeed, this is the first step in establishing autonomous control. The following actions are required:

- Clearly explain the importance of following the standard.
- Teach the skills needed for setting standards.
- Ask operators to develop and set the standards.

Operators who have been involved in initial cleaning and subsequent improvement activities will want to maintain the conditions they have established — otherwise, all their efforts will go to waste. Operators are more likely than anyone else to feel this need keenly. They should be taught the basic significance of standards through examples and given a method for defining standards by using the *5W's* and *1H*: who, what, when, where, why, and how. Thereafter, the operators can set standards based on their own experience in group meetings. Standards set in this way are certain to be obeyed.

Preparing standards enhances role-awareness. As autonomous-maintenance group members prepare their own standards, they define their own roles and make commitments to fulfill them. This is a significant developmental step. Through this process, group members begin to understand the real meaning of teamwork.

Set time targets for cleaning and lubrication and accumulating improvements. Only limited amounts of time can be spent on cleaning (including bolting and detecting minor equipment defects) and lubrication. Groups must prepare standards and individual time targets based on the limits established by management. For example, typical targets might be set at 10 minutes daily (or per shift), 30 minutes at the end of each week, and one hour at the end of each month.

If the group develops standards that cannot be completed within the targeted times, they must look for ways to reduce the times. Obviously, managers and engineering staff must cooperate wholeheartedly to simplify and improve cleaning, lubrication, and bolting procedures through measures such as centralized lubrication, extended lubrication intervals, relocating lubricators, improved lubrication instruction labels, limit marks on oil-level gauges, matching marks on bolts and nuts, use of locknuts, and various actions against contamination sources.

Example of cleaning and lubrication standards. In preparing standards, be sure to clarify the answers to the *5W's* and *1H* (although they need not all appear on the written standards) and keep in mind the concept of cleaning as inspection (Figure 4-4).

Promoting Lubrication

Ensuring proper lubrication is the second way operators can help establish basic equipment conditions. Lubrication prevents equipment deterioration and preserves its reliability. Like other hidden defects, inadequate lubrication is often neglected, because it is not always directly connected with breakdowns and quality defects.

Inadequate Lubrication Causes Losses

Losses caused by inadequate lubrication naturally include those resulting from seizures, but insufficient lubrication also leads to indirect losses such as reduced operating accuracy in sliding parts, pneumatic systems, and so on, and to accelerated wear, which hastens deterioration, produces more defects, and increases setup and adjustment times. These indirect losses can often be more significant than seizures. For example, one company found that the application of thorough lubrication control methods reduced its electricity consumption by 5 percent. Furthermore, the equipment overloading and loss in actuating accuracy produced by insufficient lubrication can readily be imagined.

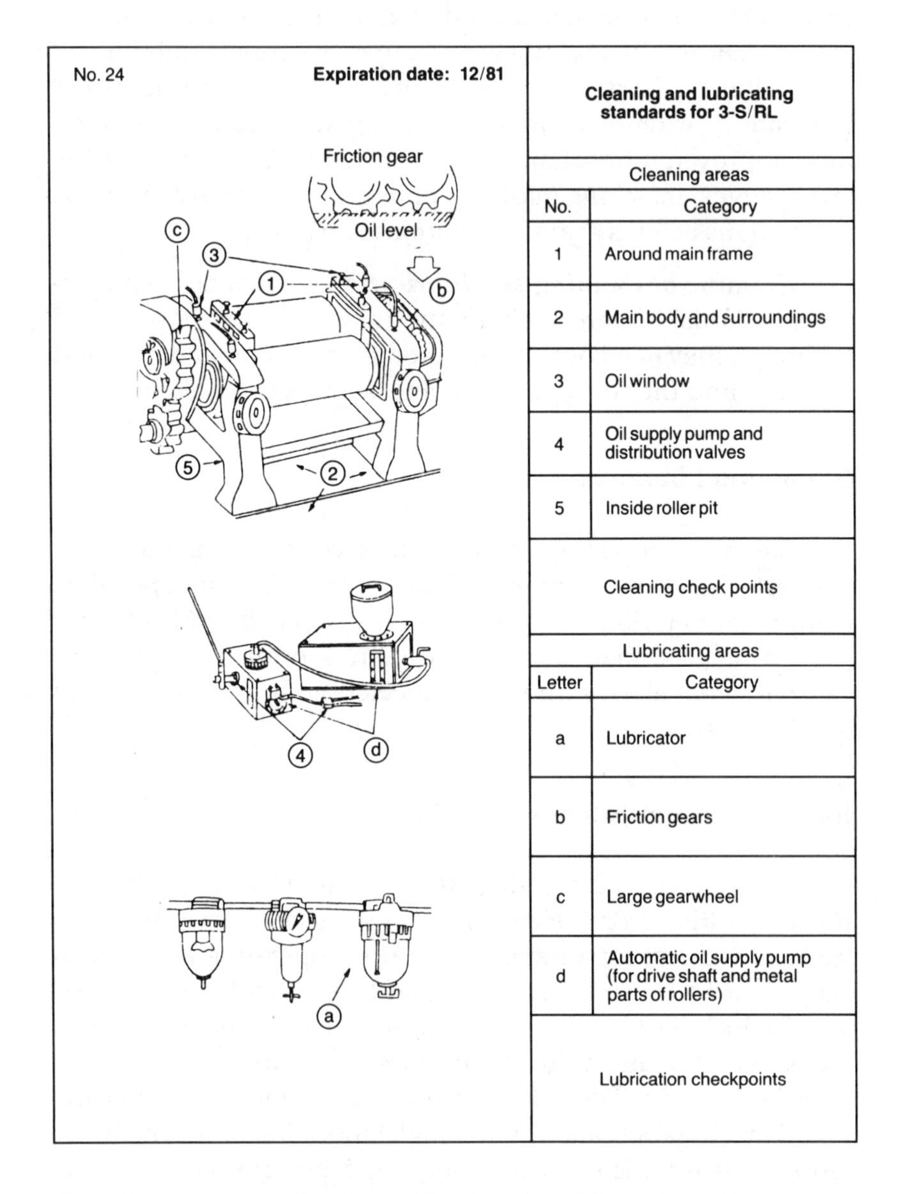

Cleaning and lubricating standards for 3-S/RL

No.	Cleaning areas: Category
1	Around main frame
2	Main body and surroundings
3	Oil window
4	Oil supply pump and distribution valves
5	Inside roller pit

Cleaning check points

Letter	Lubricating areas: Category
a	Lubricator
b	Friction gears
c	Large gearwheel
d	Automatic oil supply pump (for drive shaft and metal parts of rollers)

Lubrication checkpoints

Figure 4-4. Cleaning and Lubrication Standards (Tōkai Rubber Industries)

Cleaning and lubricating standards for 3-S/RL	Plant manager: Section chief: PM engineer: Foreman:			Raw materials plant		
Cleaning standards	Cleaning methods	Cleaning equipment	Cleaning time	Cleaning cycle		
				Day	Wk.	Mo.
No rubber scrap adhering to frame	Remove with steel scraper; sweep up		15 min.		○	
No scattering of rubber scrap	Sweep away with broom		5 min.	○		
Oil level easy to check	Wipe clean with cotton waste		3 min.		○	
No oil and dirt	Wipe clean with cotton waste		10 min.		○	
Not leaking or dirty			30 min.			○
1. Tighten the automatic supply pump ring joint bolts						
2. Tighten oil supply valve and check for leaks						
3. Tighten the stock guide fixing bolts						
Lubrication standards	Lubrication methods	Lubrication tools	Lubrication time	Lubrication cycle		
				Day	Wk.	Mo.
Oil level must be between upper and lower limit (#220)	Pour by hand		10 min.		○	
Oil level half-way up gear teeth (#32)	Oil can		5 min.			○
Gearwheel well-lubricated (open gearwheel)	Drip in through oil supply port with spatula		5 min.			○
Adequate oil (as measured by the oil gauge) (R50)	Use oil applicator		3 min.	○		
1. Secure large gearwheel cover (no rattling)						○
2. Secure attachment bolts for automatic oil supply pump						○
3. Check 3-unit FRL and drip rate					○	

Reasons for Inadequate Lubrication

The following are the most common reasons for lack of proper lubrication, apart from inadequate cleaning:

- Those performing lubrication have not been taught its basic principles or importance or shown concrete evidence of the losses caused by inadequate lubrication.
- Lubrication standards (lubrication points, types and quantities of lubricants, lubrication intervals and tools) are incomplete or not well taught.
- There are too many different types of lubricants or lubricating points.
- Not enough time is allowed for lubrication.
- Many lubrication points are inaccessible, so lubrication takes too much time.

Improving Lubrication

At one factory the maintenance manager prepared lubrication standards that actually took 30 minutes for the operator on the morning shift to perform, when only 10 minutes had been allowed. Obviously, anyone who prepares standards should test the procedure personally before giving it to the operators to make sure it can be completed in the specified time. To reduce the time, it may be necessary to make various improvements, such as changing the location of lubricators, setting up a centralized lubricating system, attaching instruction labels, and making oil levels clearly visible.

On the other hand, lubrication will serve no purpose if lubrication devices are not working or are not in good repair. A walk around a factory floor will reveal many dirty or sludgy oil reservoirs, lubricators or grease nipples, or clogged pipes in centralized lubricating systems. If this is the case, lubrication will be useless no matter how often it is done. Checkpoints for lubrication are shown in Table 4-2.

1. Are lubricant containers always capped?
2. Are lubricant stores tidy, clean, and in good order?
3. Are required lubricants always kept in stock?
4. Is all equipment labeled with lubrication instructions? Are the instructions legible?
5. Are lubricators clean inside and out and working correctly? Are oil levels always clearly visible?
6. Are all centralized lubricating systems working properly?
7. Do the reservoirs contain grease or oil, and is the system normal?
8. Are all grease and oil cups working properly?
9. After lubrication, is lubricant emerging normally from between revolving parts?
10. Is there always an oil film on revolving parts, sliding parts, and driving gears (chains, etc.); is equipment free of excess lubricant?
11. Do lubrication standards specify appropriate types and quantities of lubricants and intervals and allocation of lubrication work?

Table 4-2. Lubrication Checklist

Promoting Proper Bolting

Operators are in the best position to ensure daily that all fasteners are properly tightened. Proper bolting is the third way in which operators help establish basic equipment conditions.

Losses Caused by Improper Bolting

Loose or missing nuts, bolts, and other fastenings can cause major losses, either directly or indirectly. For example:

- Loose bolts cause fractures of dies, jigs, and tools and the production of defective products.
- Loose bolts on limit switches, dogs, and the like, and loose terminals in distribution panels and control panels cause damage and malfunctioning.
- Loose bolts in pipe flanges cause leakage.

Typically, even a single loose bolt can be the direct cause of a defect or breakdown. In most cases, however, a loose bolt causes increased vibration, producing further loosening of bolts. Vibration feeds on vibration, play feeds on play. Deterioration spreads, actuating accuracy drops, and eventually parts are damaged.

One company closely scrutinized the causes of breakdowns and found that 60 percent could be traced to faulty bolts and nuts. In another case, an inspection of all bolts and nuts revealed that out of 2,273 sets, 1,091 — an amazing 48 percent — were loose, missing, or otherwise defective.

Problems also occur frequently when dies, jigs, and tools are fastened during setup. Operators often ignore the correct torque and bolting order. Trouble results when bolts are over-tightened and when attachments are installed with immoderate force or tightened unevenly.

Assuring Proper Bolting

To eliminate loose bolts, eliminate vibration and use lock-nuts or other locking devices. In addition, place match marks on major bolts and nuts to help make loose bolts easy to spot during cleaning, and conduct inspections using test hammers.

Table 4-3 lists the checkpoints for bolts and nuts, and Figure 4-5 illustrates how to check looseness using match marks.

GENERAL INSPECTION

In an autonomous maintenance program operators are trained to conduct routine inspections. They are expected to be able to identify the often subtle evidences of deterioration.

Why Inspections Fail

Many companies ask their operators to conduct some form of inspection, but they fail to produce significant results for three familiar reasons:

- Inspection is demanded but workers are not encouraged to prevent equipment deterioration (lack of motivation through lack of direction).

- Inspection is demanded but insufficient time is allowed for it (lack of opportunity).
- Inspection is demanded but the necessary skills are not taught (lack of ability).

1. Loose bolts and nuts	No looseness
2. Proper installation of bolts and nuts	Bolts with nuts in all bolt-holes; no missing nuts
3. Use of flat washers in slots	Flat washers used in all slots (limit switch (LS) base plates, etc.)
4. Use of spring washers	No haphazard use of spring washers in similar locations
5. Loose nuts used with level-adjusting bolts	No looseness in top and bottom nuts of jack bolts holding frames in position
6. Installation of bolts and nuts	Where possible, bolts inserted from below with nut on top (as a rule, nuts to be in the most visible place)
7. Bolt lengths	Bolts long enough to allow at least 2-3 threads to emerge from nut
8. Installation of LS base plates	LS base plates to be fixed with at least 2 bolts
9. Other	Special items other than the above

Action

In principle, TPM groups should handle defects that do not require machining (submit work orders to the maintenance department for removal of broken-off studs, screw-tapping, etc.)

- Mark faulty bolts and nuts and repair them right away if possible
- Identify those that might require equipment to be stopped, (*e.g.*, cannot be repaired while equipment is running or unsafe to handle). Decide how to deal with these after discussing with supervisor
- Identify items the group cannot handle. Submit work orders for these as part of improvement list

⋆ Check marks: if OK, mark in white. Mark loose bolts and nuts or places lacking bolts and nuts in yellow

Table 4-3. Checklist for Bolts and Nuts

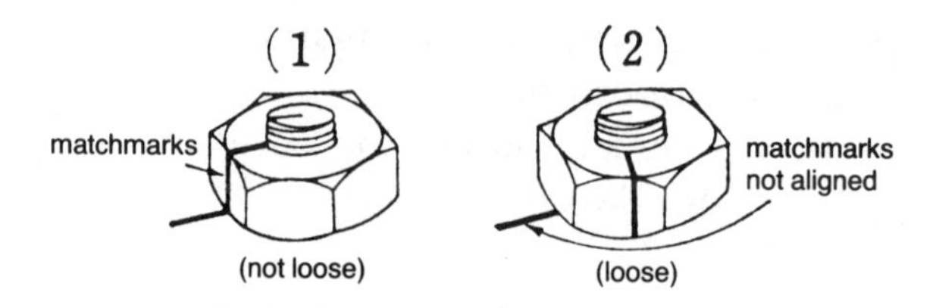

Figure 4-5. Checking for Looseness with Matchmarks

Inspection Functions Are Not Well-Defined

Problems with inspection are inevitable when maintenance engineers prepare inspection checksheets and simply hand them to the operators. The engineers invariably want too many items inspected and tend to consider their job finished when they have prepared the checksheets. They do not indicate which items are most important to check and how much time is needed; nor do they consider that inspection procedures might be streamlined or that operators may need to be taught certain skills in order to perform them.

Operators Need Inspection Skills

Asking operators to perform inspections for which they do not have the necessary skills is probably the biggest problem. Typically, operators are asked to perform visual inspections, but these are often quite difficult because deterioration cannot be identified or measured visually in many cases. Operators need considerable training before they can perform inspections properly; simply handing them a checksheet does not enable them to do the job.

Even trained operators should not rely exclusively on checksheets to perform inspections. Their most important function is to make judgments based on their knowledge of the ongoing condition of their equipment and to identify malfunctions during cleaning and lubricating. Line managers and engineering staff must realize that checksheets cover only a small portion of the operators' work. (For example, it is impossible to make a checksheet listing every single nut and bolt to be inspected!)

The first requirement for autonomous general inspection is operators who are knowledgeable and confident about their equipment. Once operators have had instruction in inspection skills and practice in conducting general inspections, they can prepare checksheets that meet their own requirements.

Determining inspection intervals and times is also critical, since the work must be done while equipment is operating.

Inspection Intervals

Suitable inspection intervals for autonomous maintenance can be daily, every ten days (or bi-weekly), monthly, and every three months. Very little time can be spared for daily inspection on the production line because considerable time is already spent daily preparing for startup, clearing up after shutting down, as well as in cleaning and lubrication. Thus, daily inspections should focus only on equipment deterioration directly affecting safety and product quality.

Keep Daily Inspections Simple

Many factories ask their operators to follow a very detailed inspection procedure (*i.e.*, checklists with many items). Often, however, many of the items listed need not be checked every day. In addition, insufficient time is allowed. This only antagonizes operators and renders the checklists meaningless.

Limit daily inspections to the few items necessary to prevent serious safety and quality problems and let operators practice them thoroughly until they become part of each individual's routine. Avoid putting the items on a checklist — if they must be listed, there are probably too many.

Allow Adequate Time for General Inspection

General inspection is far too important to be carried out hastily and haphazardly in daily procedures loaded with inessential items. Consider, instead, setting aside a block of time to devote full attention to it, even if the interval between inspections must be extended. (For example, set aside an extra fifty minutes every ten days in place of five minutes a day.)

In an extended block of time, operators are sure to perform all the required checks thoroughly and reliably, with time to spare. Moreover, as they practice this activity, operators gradually develop the ability to read the condition of their equipment during operation and to detect signs of trouble while cleaning or lubricating — even without setting aside special time.

Establish Realistic Inspection Intervals

The interval for each inspection item can be determined only through experience. Taking into consideration the time restrictions, the responsible production and maintenance staff should agree on a suitable time based on their own experience and on the likelihood and possible consequences of equipment breakdown. Of course, inspection intervals are subject to modification in light of the subsequent inspection record. Figure 4-6 is an example of inspection intervals at company C.

Inspection Times

Whether to inspect and how much time can be spent on each inspection item depends on the equipment and its environment as well as on the inspection interval. This question is affected by factors such as whether the operator merely supervises the equipment or is continually working on it (the degree of automation), whether the equipment is critical to the speed of the production line, whether most of the inspection can be performed while the equipment is running, and so on.

In most cases, inspections are limited by the amount of time that can be spared, so it is best to start by considering the factors mentioned above and work out temporary time targets as a guideline. Determine inspection items and intervals in advance and prepare checksheets for each interval. Carry out inspections following these checksheets and record the differences between the

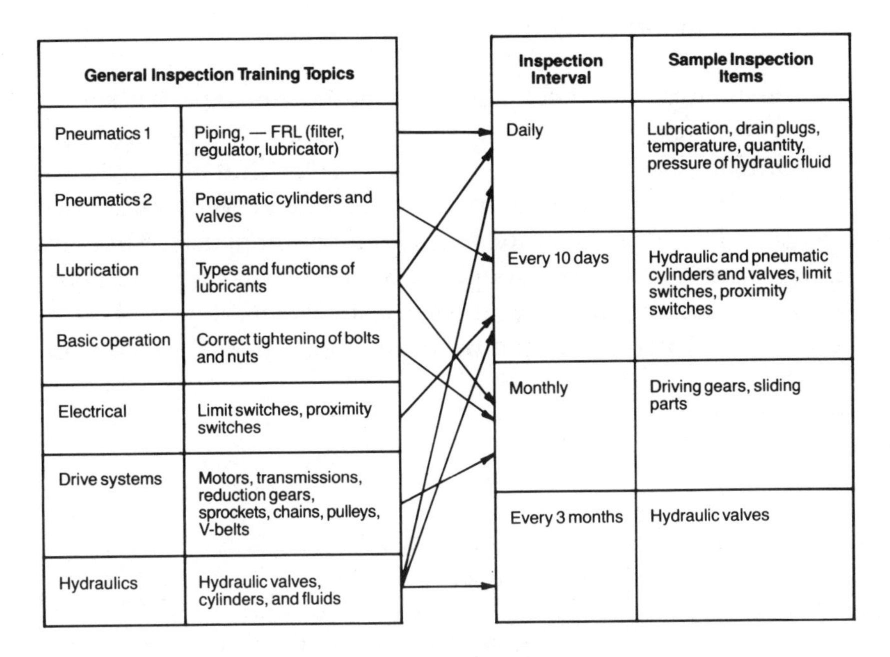

Figure 4-6. Relation Between Inspection Training and Periodic Inspection Items

time targets and the times actually taken. At this point it may be necessary to simplify inspection procedures and reduce cleaning and lubricating times; inspection intervals and allocations will also have to be reconsidered. Although certain inspection items may take time at first, times will be considerably reduced as operators become familiar with the procedures. Figure 4-7 outlines inspection times at company C.

Interval	Month 1	Month 2	Month 3	Time required
Daily				Approx. 10 min.
Every 10 days				15-20 min.
Monthly				15-20 min.
Every 3 months				15-20 min.

Figure 4-7. Required Inspection Times

Checkpoints for Daily Inspections

In determining checkpoints for daily inspections, it is convenient to consider the units common to most machines. The pneumatic, steam, hydraulic, drive, and electrical units referred to here can be found on almost all automated equipment. Since they break down frequently, they should be emphasized in daily inspections. Tables 4-4 through 4-7 are checklists for pneumatic, steam, hydraulic, drive, and electrical systems. Figure 4-8 (*see* pp. 197-198) is an example of autonomous maintenance standards.

1. Pneumatics	a. Are FRLs being used correctly? b. Are solenoid valves over-heating? Check for loose and damaged wiring. c. Is there slackness in air cylinders, foreign matter on cylinder rods, or damage to cylinder rods? d. Are speed controllers correctly installed (in flow direction)? e. Is any pneumatic equipment leaking?
2. Piping and Equipment	a. Are there any loose bolts, vibration, or bent pipes? b. Are there any steam, air, or water leaks, or leaking steam drains? c. Has any discarded piping been left in place? d. Are there any loose hose clips or joints?
3. Valves and Insulation	a. Are there any damaged valves, missing handwheels, stopcocks, or loose bolts? b. Do all valves shut the flow off completely when closed? c. Do handwheels turn easily? Are they difficult to open or close? d. Are all steam and air pressure gauges clean and undamaged? Do they have maximum and minimum marks? e. Is any piping or equipment insulation hanging loose or torn?
4. Inspections and Inspection Standards	a. Have efforts been made to make inspecting easier? b. Are inspection frequencies, intervals, and allocations appropriate for autonomous maintenance? c. Do the inspection standards take into account safety, breakdowns, and product quality?

Table 4-4. Pneumatics/Steam Checklist

1. Hydraulic Units	a. Do reservoirs contain the specified quantities of hydraulic fluid? Are maximum and minimum levels displayed? b. How hot is the fluid in the reservoirs? Can it be touched? c. Is the reservoir cooling water flowing properly? d. Are filters unclogged? Are indicators blue? e. Are pressure-gauge zero points accurate? Are needles free from deviation? Are maximum and minimum marks displayed? f. Are there any unusual noises or smells? g. Is there any play in devices or piping? Is there any leaking fluid? h. Are any units contaminated by water, oil, dust, or other foreign matter? i. Are the name plates on each device easy to read?
2. Piping and High-Pressure Hoses	a. Are there any leaking joints or hoses? b. Is there any looseness or play in clamps? c. Are piping pits free of spilt fluid? d. Are any high-pressure hoses dirty or damaged?
3. Hydraulic Equipment	a. Is equipment free of damage (covers, lids, etc.?) b. Is equipment securely installed, without play or looseness? c. Are pressure gauge needles correctly zeroed and free from deviation? d. Is equipment actuating correctly? (speed, breathing, chatter) e. Are pressure gauges calibrated regularly and registered in the instruments store?
4. Presses	a. Are press stroke speeds the same as usual? b. Are relief valves set to the correct pressures? c. Are relief-valve spring-adjusting screw locknuts securely tightened?
5. Inspection Standards	a. Are inspection intervals and work allocations appropriate for autonomous maintenance? b. Do the standards take into account safety, breakdowns, and product quality?

Table 4-5. Hydraulics Checklist

AUTONOMOUS INSPECTION TRAINING AND EDUCATION

Educating operators about their equipment is cost-effective but time-consuming because training must be detailed, and it must begin with the basics (*e.g.*, equipment functions, mechanisms, and operating principles). Moreover, operators should be trained on the equipment they use, which complicates production scheduling. Many companies are discovering, however, that thorough technical training is the key to establishing TPM and obtaining significant benefits. The training programs described in this section reflect this understanding.

1. V-Belts	a. Are any surfaces damaged, split, contaminated by oil, or badly worn? b. Is tension of multiple V-belts uniform? c. Are any non-standard belts in use?
2. Roller Chains	a. Is lubricant penetrating between pins and bushes? b. Is there poor meshing because of stretched chains or worn sprockets?
3. Shafts, Bearings, Keys and Couplings	a. Are there any overheated, vibrating, or noisy bearings due to bent or off-center shafts, loose bolts, lack of lubrication, etc? b. Is there play in bosses due to loose keys or set screws? c. Are there any deflected flange coupling shafts or loose bolts?
4. Gears, Speed Reducers, and Brakes	a. Are there any noisy, vibrating, or unusually worn gears? b. Are oil gauges provided with maximum and minimum marks? Are all oil levels between the marks? c. Are brakes operating correctly? d. Are any safety covers in contact with rotating parts?
5. Inspection Standards	a. Are inspection frequencies, intervals, and work allocations appropriate for autonomous maintenance? b. Do the standards take into account safety, breakdowns, and product quality?

Table 4-6. Checklist for Drive Systems

Developing Operators Who Understand Their Equipment

Since operators use their equipment to make products, they might be expected to be reasonably familiar with it. Managers in many companies believe, however, that their operators merely need to know how to fit and remove appropriate attachments and follow instructions. At such companies, no effort is made to develop operators' understanding of their equipment. On the other hand, other companies are finding that increasing operators' knowledge and skills can transform the way a factory or shop floor is run and produce outstanding benefits.

A knowledgeable operator does not need the repair skills of a maintenance worker. Rather, the operator's most important skill is the ability to spot abnormalities. Operators must know enough about the equipment to pinpoint small signs of trouble whenever anything out of the ordinary occurs.

1. Wiring	a. Are all wiring, conduit piping, and flexible connectors securely fixed? b. Are all grounding cables securely fixed? c. Is any vinyl or rubber-sheathed cable loose on walkways or damaged?
2. Control Panels	a. Is there any deviation in voltmeters, ammeters, thermometers, or other instruments? b. Are there any burnt-out bulbs in pilot lamps and display lamps? c. Are push-button switches and other types of switch firmly fixed? d. Are there any unnecessary holes? Does the door open and close properly? e. Is the wiring inside panel boxes tidy? f. Are the interiors of panel boxes free from dust, dirt, etc.? g. Do panel boxes contain anything other than diagrams?
3. Electrical Equipment	a. Is any equipment damaged? Are any motors overheating? b. Are any bolts loose? c. Are there any unusual noises or smells? Are bearings well lubricated? d. Are heaters securely fixed? e. Are all grounding cables securely fixed? f. Are limit switches, proximity switches, and photoelectric tubes clean and free from play in main body and fixing bolts? g. Are wires to equipment in contact with steam, oil, or water? h. Is all electrical equipment free of water, oil, dust, and other foreign matter?
Inspection Standards	Are inspection frequencies, intervals, and work allocations appropriate for autonomous maintenance? Do the standards take into account safety, breakdowns, and product quality?

Table 4-7. Electrical Checklist

Operators Should Be Able to Detect Causal Abnormalities

The word *abnormality* in this context does not refer to abnormal *effects*. If a machine breaks down and stops or produces defective products, these are effects that every operator must be able to recognize. More difficult to recognize, however, are *causal abnormalities* — conditions that can lead to a breakdown or result in defective products. These abnormalities manifest *before* breakdowns or defects occur, at points where they can be prevented. A truly skilled operator can detect causal abnormalities and deal with them promptly.

Developing operators with these skills is not easy, but it is essential as long as we need equipment to make products. Such operators not only change the way equipment is managed, they make possible sweeping changes in all shop-floor management practices. The autonomous inspection skills and training discussed here should be regarded as the first step in developing knowledgeable and skilled operators — operators who can rightly be called "human sensors."

Preparing for General Inspection Training

If we accept the challenge of developing operators with these skills, we must train them well (*see* Figure 4-9; p. 199). An important part of that training prepares operators to conduct general inspection.

Identify General Inspection Items

The curriculum to be taught will depend on what operators need to know — for example, how to set conditions, how to set up and operate equipment correctly, and how to conduct a thorough inspection. This is determined in accordance with the design specifications of the particular equipment and the incidence of breakdowns, defects, and other problems. The program should also cover (at least) the basic functional components of the equipment (*e.g.*, bolts and nuts, lubrication, pneumatics, hydraulics, drive system, electricity, and instrumentation).

Prepare Checksheets and Manuals

The most important training materials are the general inspection checksheets and manuals. First, sort out the items that operators should inspect using the senses and incorporate them in general inspection checksheets.

Then consider what technical information operators will need to know to master the skills required to inspect the items. Include this information in inspection manuals aimed at group leaders. Be sure to provide relevant details such as the basic functions, mechanisms, and components of the units to be inspected,

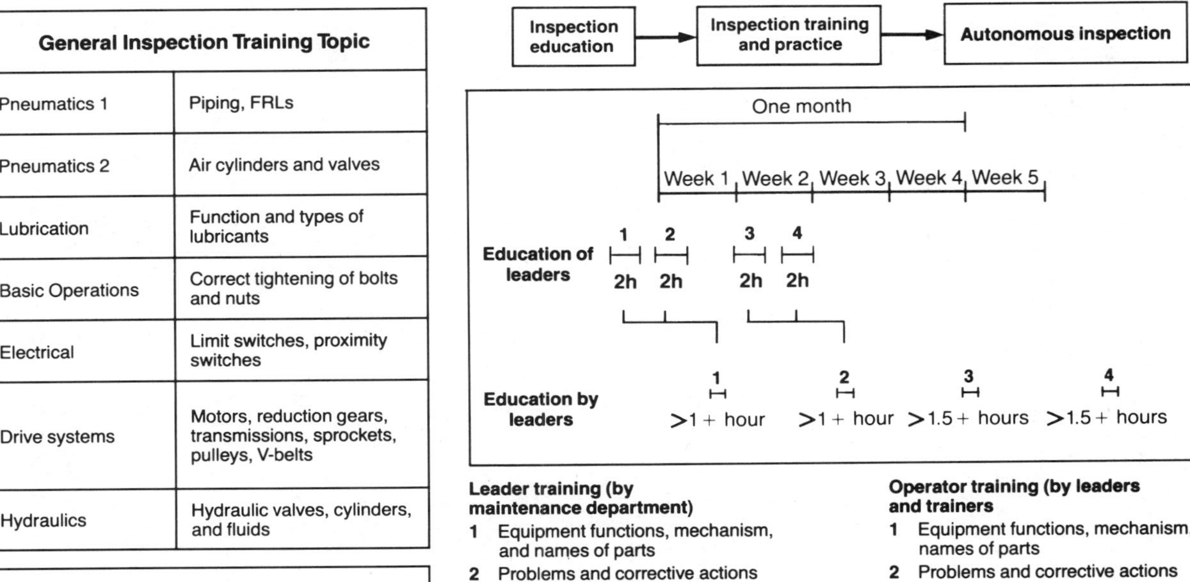

General Inspection Training Topic	
Pneumatics 1	Piping, FRLs
Pneumatics 2	Air cylinders and valves
Lubrication	Function and types of lubricants
Basic Operations	Correct tightening of bolts and nuts
Electrical	Limit switches, proximity switches
Drive systems	Motors, reduction gears, transmissions, sprockets, pulleys, V-belts
Hydraulics	Hydraulic valves, cylinders, and fluids

Equipment-specific topics

Leader training (by maintenance department)

1 Equipment functions, mechanism, and names of parts
2 Problems and corrective actions
3 Inspection priorities, methods, criteria, etc.
4 Inspection practice and review

Items to be carried out

Pneumatics 1, pneumatics 2, lubrication, basic operations, electrical, drive systems, hydraulics: 1 item per month

Operator training (by leaders and trainers

1 Equipment functions, mechanism, and names of parts
2 Problems and corrective actions
3 Inspection training and review (on-the-job training, meetings)
4 Autonomous inspection and review (on-the-job training, meetings)

Figure 4-10. Sample Schedule for General Inspection Training Topics

The rationale behind this group-based approach is not to spare the maintenance staff from having to teach all the operators directly. Rather it fosters leadership skills in group leaders and team spirit within groups by locating the learning process in the group environment. In educating their own group members, leaders learn the responsibilities of leadership. Assuming responsibility for group education forces leaders to take their own skill development seriously. Group members feel their leaders' enthusiasm, sympathize with their burden of responsibility, and make a greater effort to cooperate. The TPM groups become even more active as a result of this teaching and learning process.

Group Leaders Must Learn How to Teach

The maintenance staff instructors must do more than ensure that leaders understand the content of their instruction. If the group-based educational process is to succeed, they must also teach the leaders *how* to teach — how to present information effectively, using wall charts or other visual aids to clarify important points.

Leaders Must Prepare for Group-Based Education

The leaders must also do more than simply pass on what they are taught. Ideally, they should plan their training sessions by discussing important points with their own managers and preparing training materials that are keyed to the equipment installed in their group's work area.

Conduct Training Outside the Classroom

Group-based education should not be confined to the classroom. Meetings should be held near equipment whenever possible, to let group members examine their own equipment during instruction and to let leaders answer questions using the equipment itself.

Make Learning Enjoyable

Everyone learns more when the training is enjoyable. Have group members actually dismantle simple units or ask them to do case studies of problems actually occurring on the shop floor. Introduce an element of friendly rivalry by dividing the members into groups and having them compete against each other to find equipment defects.

Conducting General Inspections

The goal of inspection education is to give operator groups the ability to inspect and restore their equipment. The groups use the knowledge they have gained to conduct general inspections of all their equipment. Then in their meetings they discuss the results, deal with the malfunctions they have discovered, and develop ways to inspect less accessible areas more easily. Practicing this type of activity over time gradually increases group members' inspection skills and improves the reliability of their equipment.

During this activity, the cooperation of the maintenance department is crucial. At this stage in the TPM development process, operators discover a considerable amount of deterioration. More than half of the malfunctions found will have to be restored by the maintenance department, which will be flooded with work requests. If the work cannot be carried out promptly, operator groups' morale will drop and the inspection training may fail.

The maintenance department must therefore do everything in its power to respond to these work orders while carrying out its standard workload.

Completing the General Inspection Step by Step

As each general inspection item is completed, groups should consider what kind of inspection routine will be required to maintain the improved state of the equipment and prepare preliminary autonomous inspection standards. At the same time, the leaders should assess group members' inspection skills and give further training in weak areas.

Management groups should audit the autonomous maintenance program after the completion of each general inspection item. They should assess the extent of improvement by measuring the results of inspections on the shop floor, point out any problems, and give further guidance and encouragement.

SEIRI AND *SEITON* — KEYS TO WORKPLACE MANAGEMENT

Seiri (organization) and *seiton* (tidiness) are fundamental principles of workplace management. They are easy to promote but notoriously hard to put into practice. Although many companies have posted these concepts in large letters on their factory walls, most still find it hard to instill the good habits the words imply. Doing so requires role-awareness and unrelenting improvement efforts based on the attitudes described below.

Seiri and *Seiton* = Standardization

Seiri (organization), introduced earlier in Chapters 1 and 3, involves the identification of objectives to be managed and the setting of relevant standards, for example, height standards for stacking containers. One object, therefore, is to minimize the number of items or conditions to be managed and to simplify them as much as possible. Managers and supervisors are responsible for guiding this activity.

Seiton (tidiness) refers to adherence to set standards. This is primarily the operators' responsibility. Since they must ensure adherence to standards, group activity regularly focuses on improvements that make the standards easier to obey, typically through the creative use of visual controls.

Seiri and *seiton* are thus improvement activities designed to simplify and standardize what must be organized and controlled and to find creative ways to improve adherence. In this way, managers and workers cooperate to ensure that standardization and visual control reach into every part of the factory.

Seiri and the Operators' Role

In addition to maintaining basic equipment conditions and inspecting the equipment, the operators' role in autonomous maintenance includes the following:

- correct operation and setup/adjustment (setting operation conditions and checking product quality)
- rapid detection and prompt, reliable treatment and reporting of abnormal conditions (breakdowns, quality defects, safety, etc.)
- recording data on operation, quality, and processing conditions
- minor servicing of machines, dies, jigs, and tools
- control of any other items required to reliably accomplish the above

To improve the standardization process, managers and supervisors must take into account current losses due to breakdowns, defective products, and so on, and take the lead in answering the following questions:

- What are operators required to do?
- Are operators performing well in all areas?
- If they are not performing well, what are the reasons?
- What standards must be set?
- What skills do operators lack?
- What can be done to increase operators' skills?

Organization and Management of Materials and Tools

In addition to machinery, a factory must manage large numbers of dies, jigs, tools, work-in-process, finished products, defective products, measuring instruments, material-handling equipment, auxiliary apparatus, ancillary materials, and so on.

To eliminate losses and to detect and deal with abnormalities rapidly, all these items must be organized and maintained optimally in terms of both quality and quantity.

Ideally, everything we need should be where we want it, when we want it, at precisely the time we want it, in exactly the right quantity, and fully equipped with the required functions. To achieve this the following conditions must be fulfilled:

- Decisions have been made as to what will be used, when it will be used, who will use it, and how much will be used.
- Both quality (can it perform the required functions well?) and quantity have been checked.
- Items have been assigned locations and quantity levels that can be seen at a glance; they are available when needed without searching.
- Locations and methods of storage have been selected according to frequency of use, so that items occupy the minimum space and can easily be removed and transported to where they are needed.
- Appropriate individuals have been put in charge of routine control, and methods of supply and disposal have been established.

Table 4-8 is an example of a workplace management (*seiri* and *seiton*) program.

ASSURING CORRECT OPERATION

Operators must operate equipment correctly, and most companies have "work procedures," "operating standards," or other similar manuals that specify operating methods. Seriously asking the question, "What *is* correct operation?" will usually expose a variety of problems, however.

Progress in hydraulics, pneumatics, microelectronics, and instrumentation has made equipment more sophisticated and complex. Some equipment has become more difficult to operate and some has become more simple. In either case, the damage caused by misoperation has become immeasurably greater than

Focus	Elements
Operator's responsibility	Organize standards for operator responsibilities; adhere to them faithfully (including data recording)
Work	Promote organized and tidy operations as well as visual control of work-in-process, products, defects, waste, and consumables (such as paint)
Dies, jigs and tools	Keep dies, jigs, and tools organized and easy to find through visual control; establish standards for precision and repair
Gauges and defect prevention devices	Inventory gauges and defect prevention devices and make sure they function properly; conduct general inspection and correct deterioration; set standards for inspection
Equipment precision	Operators must check precision of equipment (as it influences quality) and standardize procedures
Operation and treatment of abnormalities	Establish and monitor operation, setup/adjustment, and processing conditions; standardize quality checks; improve problem-solving skills

Table 4-8. Sample Workplace Organization and Housekeeping Program

it was when machine mechanisms were simple. Thus we must teach operators *why* equipment must be operated in the specified way, by explaining the equipment's construction, mechanism, and functions, and the principles behind the processing of a product or the chemical changes it undergoes. We must also train operators to operate their equipment efficiently and correctly in any situation.

Unfortunately, this instruction is often inadequate, and many operators do not understand their equipment well enough to avoid causing equipment failures and accidents.

Table 4-9 lists the autonomous-maintenance checkpoints concerned with correct operation.

1. Machine Operation	a. Are operating procedures and pre- and post-operations set down? Are they taught and observed? • Startup • Setting conditions • Adjusting conditions • Cycle modification • Emergency shutdown • Routine shutdown b. Have efforts been made to simplify operation and minimize the possibility of error? c. Is information on problems caused by operating errors gathered and fully used? d. Has it been decided when and how adjustments should be made? e. Can equipment be started by mistake even when it should be impossible for safety reasons? f. Are operating procedures to be taught to new operators clearly set down? g. Are operators' positions correct?
2. Dealing with Problems	a. Are there clear rules for reporting and dealing with operating errors? b. Are abnormal conditions clearly defined? c. Have efforts been made to make abnormal conditions easy to detect visually? d. Are there definite procedures for dealing with problems? Are they followed? e. Are operators taught and encouraged to practice the above procedures? f. Are operators actually detecting problems in equipment, product quality, and safety?
3. Equipment Functions	a. Is there any looseness or play in levers and wheels? Are they hard to operate? b. Are controls and other parts requiring manipulation accessible and well lit? c. Does equipment start and stop correctly? d. Are all measuring instruments operating properly? Are maxima and minima shown? e. Are emergency shutdown devices operating correctly? f. Is there any abnormal noise, heat, or vibration? g. Are valves marked with functions and open/close arrows?

Table 4-9. Autonomous Maintenance Checklist for Machine Operation

IMPLEMENTING AUTONOMOUS MAINTENANCE IN SEVEN STEPS

Table 4-10 (*see* pp. 210-211) outlines the seven developmental stages of an autonomous maintenance program. These stages or steps are based on the experiences of many companies that have successfully implemented TPM. They represent an optimal division of responsibilities between production and maintenance departments in carrying out maintenance and improvement activities.

Each stage in the implementation of autonomous maintenance emphasizes different developmental activities and goals, and each builds upon thorough understanding and practice of the previous step. Step 1 (*initial cleaning*), step 2 (*action against the sources of dust and contamination*), and step 3 (*cleaning and lubrication standards*) promote the establishment of basic equipment conditions that are essential to effective autonomous maintenance.

Step 4 (*general inspection*) and step 5 (*autonomous inspection*) stress thorough equipment inspection and subsequent maintenance and standardization. Furthermore, these steps promote the development of operators who are knowledgeable and sensitive to their equipment's needs. During these periods, the company is likely to see substantial reductions in equipment failures.

Steps 6 and 7 (*seiri and seiton* and *full autonomous maintenance*) stress improvement activities informed by operators' increasing knowledge and experience and extending beyond the equipment to its surrounding environment. These activities increase the vitality of operators' involvement as well as the skills they acquired in the earlier stages. Operators become strongly identified with company goals and assume responsibility for the maintenance and improvement activities that are essential for effective self-management on the shop floor.

KEYS TO SUCCESSFUL AUTONOMOUS MAINTENANCE

To implement autonomous maintenance successfully, consider the following important elements:

Introductory Education and Training

Before initiating any of the early autonomous maintenance activities, make certain all related departments and personnel (from top management to shop-floor supervisors) understand the objectives and benefits of TPM development. Require everyone to attend an introductory seminar that spells out the details of TPM implementation and, in particular, the functions of autonomous maintenance.

Step	Activity	Goals for Equipment (workplace diagnosis)
1. Initial Cleaning	Thoroughly remove dust and contaminants from equipment (remove discarded equipment parts)	• Eliminate environmental causes of deterioration such as dust and dirt; prevent accelerated deterioration • Eliminate dust and dirt; improve quality of inspection and repairs and reduce time required • Discover and treat hidden defects
2. Eliminate Sources of Contamination and Inaccessible Areas	Eliminate the sources of dust and dirt; improve accessibility of areas that are hard to clean and lubricate; reduce time required for lubrication and cleaning	• Increase inherent reliability of equipment by preventing dust and other contaminants from adhering and accumulating • Enhance maintainability by improving cleaning and lubricating
3. Cleaning and Lubrication Standards	Set clear cleaning, lubrication, and bolting standards that can be easily maintained over short intervals; the time allowed for daily/periodic work must be clearly specified	• Maintain basic equipment conditions (deterioration-preventing activities): cleaning, lubrication, and bolting
4. General Inspection	Conduct training on inspection skills in accordance with inspection manuals; find and correct minor defects through general inspections; modify equipment to facilitate inspection	• Visually inspect major parts of the equipment; restore deterioration; enhance reliability • Facilitate inspection through innovative methods, such as serial number plates, colored instruction labels, thermotape gauges, and indicators, etc.
5. Autonomous Inspection	Develop and use autonomous maintenance checksheet (standardize cleaning, lubrication, and inspection standards for ease of application)	• Maintain optimal equipment conditions once deterioration is restored through general inspection • Use innovative visual control systems to make cleaning/lubrication/inspection more effective • Review equipment and human factors; clarify abnormal conditions • Implement improvements to make operation easier
6. Workplace Organization and Housekeeping (workplace management and control)	Standardize various workplace regulations; improve work effectiveness, product quality, and the safety of the environment: • Reduce setup and adjustment time; eliminate work-in-process • Material handling standards on the shop floor • Collecting and recording data; standardization • Control standards and procedures for raw materials, work-in-process, products, spare parts, dies, jigs, and tools	• Review and improve plant layout, etc. • Standardize control of work-in-process, defective products, dies, jigs, tools, measuring instruments, material handling equipment, aisles, etc. • Implement visual control systems throughout the workplace
7. Fully Implemented Autonomous Maintenance Program	Develop company goals; engage in continuous improvement activities; improve equipment based on careful recording and regular analysis of MTBF	• Collect and analyze various types of data; improve equipment to increase reliability, maintainability, and ease of operation • Pinpoint weaknesses in equipment based on analysis of data, implement improvement plans to lengthen equipment life span and inspection cycles

Table 4-10. Developing Autonomous Maintenance Small-Group Activities

Goals for Group Members (TPM group diagnosis)	Management's Leadership
• Develop curiosity, interest, pride, and care for equipment through frequent contact • Develop leadership skills through small group activities	• Teach control of dust and dirt, equipment deterioration, and related maintenance work • Identify priority areas to be cleaned and the importance of the maintaining basic equipment conditions (according to cleaning, lubrication, and bolting standards) • Teach what "cleaning is inspection" means
• Learn equipment improvement concepts and techniques, while implementing small-scale improvements • Learn to participate in improvement through small-group activity • Experience the satisfaction of successful improvements	• Offer guidance that is easy to understand in response to operators' questions • Make sure maintenance work orders are carried out promptly • Promote visual control systems (*e.g.*, use of lubrication labels and match marks, etc.)
• Understand the meaning and importance of maintenance by setting and maintaining our own standards (What is equipment control?) • Become better team members by taking on more responsibility individually	• Provide guidance on the content and form of cleaning standards • Provide technical assistance in the development of lubrication standards
• Learn equipment mechanisms, functions, and inspection criteria through inspection training; master inspection skills • Learn to perform simple repairs • Leaders enhance leadership skills through teaching; group members learn through participation • Sort out and study general inspection data; understand the importance of analyzed data	• Prepare general inspection manuals and problem case studies; provide inspection training to group leaders • Prepare schedules for general inspection • Provide prompt action against work orders issued through general inspection • Teach simple treatment for minor defects • Simplify inspections through creative use of visual controls • Provide instruction in data collection and analysis • Invite group leaders to participate in planned maintenance scheduling
• Draw up individual daily and periodic checksheets based on general inspection manual and equipment data and develop autonomous management skills • Learn importance of basic data-recording • Learn proper operating methods, signs of abnormality, and appropriate corrective actions	• Provide guidance in developing inspection priorities and intervals based on data analysis • Give advice on the content and form of inspection checksheets • Provide technical assistance in developing operation standards and troubleshooting manuals
• Broaden the scope of autonomous maintenance by standardizing various management and control items • Be conscious of the need to improve standards and procedures continuously, based on a standardization practice and actual data analysis • Managers and supervisors are primarily responsible for continuously improving standards and procedures and promoting them on the shop floor	• Provide technical assistance as needed to TPM groups and departments • Teach improvement techniques, visual control systems, IE, and QC methods
• Gain heightened awareness of company goals and costs (especially maintenance costs) • Learn to perform simple repairs through training on repair techniques • Learn data collection and analysis and improvement techniques	• Provide technical assistance for equipment improvement • Provide training in repair techniques • Participate in equipment improvement meetings; encourage groups to strive for continuous improvement • Standardize improvement results

Cooperation Between Departments

The managers of all relevant departments (production, maintenance, engineering, design, personnel, general affairs, and accounting, etc.) must meet and agree on how the departments will cooperate to support the efforts of the production department to achieve autonomous maintenance.

Group Activities

Most activities are performed in small groups in which all personnel participate. Group leaders at every level are part of the company management structure. For example, TPM groups at the shop-floor level are organized around the production supervisors. If there are too many group members, the group can be divided into sub- or mini-groups of approximately five to ten members. The leaders of these groups form a group themselves under the leadership of a shift superintendent, who in turn works in a group under a department manager. These managers also participate in a group lead by the plant manager, who is part of a TPM promotional group. This group may be composed of other plant managers and division managers.

Thus, the TPM promotional structure is organized along the lines of the company hierarchy. Group leaders at each level become members of upper-level small groups and serve as links between levels.

To manage such a promotional structure effectively may require an administrative office for the TPM promotional committee. PM administrators may be assigned as needed to special committees or individual groups at any level to provide additional support.

Autonomous Maintenance Is Not a Voluntary Activity

All participants must understand that autonomous maintenance activities (at all stages) are mandatory and necessary. Some managers and supervisors imagine "autonomous" maintenance to be synonymous with "voluntary" or "unregulated and unchecked" maintenance. If this misunderstanding is not corrected, it can undermine group activities. The autonomous maintenance work done by operators is designed to support and complement the maintenance department's planned maintenance effort. Once begun, the activities of both departments must continue side by side.

Moreover, these activities are a vital, necessary part of daily work. Their purpose is to achieve company goals through the implementation of operator-initiated daily maintenance consisting of cleaning, adjustment, and regular inspections, as well as improvement activities.

Autonomous maintenance requires skilled, highly motivated operators working in a supportive, rational work environment. Management must provide workers with the leadership and guidance they need to enhance their skills and motivation and must cooperate at every level to maintain a favorable working environment.

Practice

Understanding comes through practice rather than intellectual rationalization. Use the activities themselves for primary instruction to avoid being misled by purely conceptual manipulation or empty exercises in logic.

Education and Training Should Be Progressive

Successful implementation of autonomous maintenance depends on a combination of gradual skill development, experiential learning, and expanding awareness or attitudinal change. Each step in autonomous maintenance builds on the knowledge, experience, and understanding acquired in the previous step. For this reason, education and training must be designed to progress step by step, and must be carefully tailored to meet the changing needs of operators and the manufacturing environment.

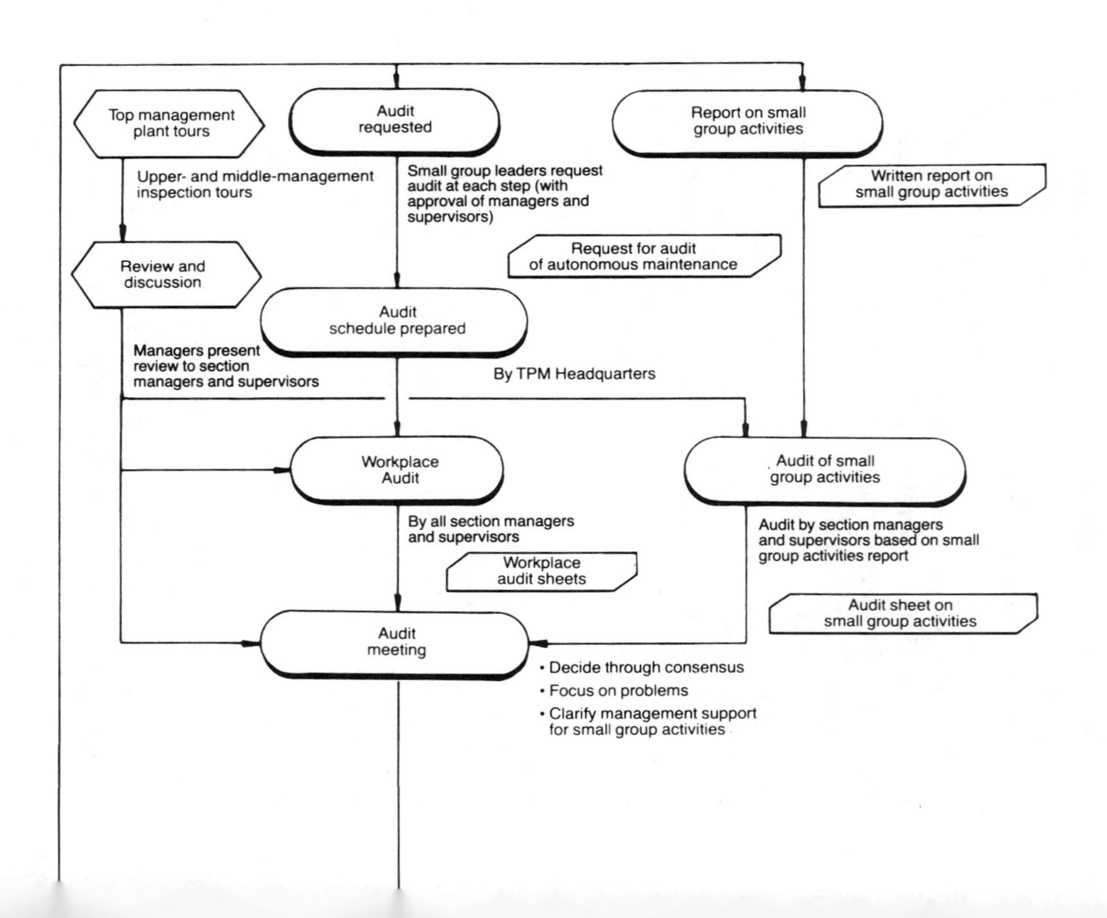

Top management plant tours
Upper- and middle-management inspection tours
Review and discussion
Managers present review to section managers and supervisors
Audit requested
Small group leaders request audit at each step (with approval of managers and supervisors)
Request for audit of autonomous maintenance
Audit schedule prepared
By TPM Headquarters
Workplace Audit
By all section managers and supervisors
Workplace audit sheets
Audit meeting
• Decide through consensus
• Focus on problems
• Clarify management support for small group activities
Report on small group activities
Written report on small group activities
Audit of small group activities
Audit by section managers and supervisors based on small group activities report
Audit sheet on small group activities

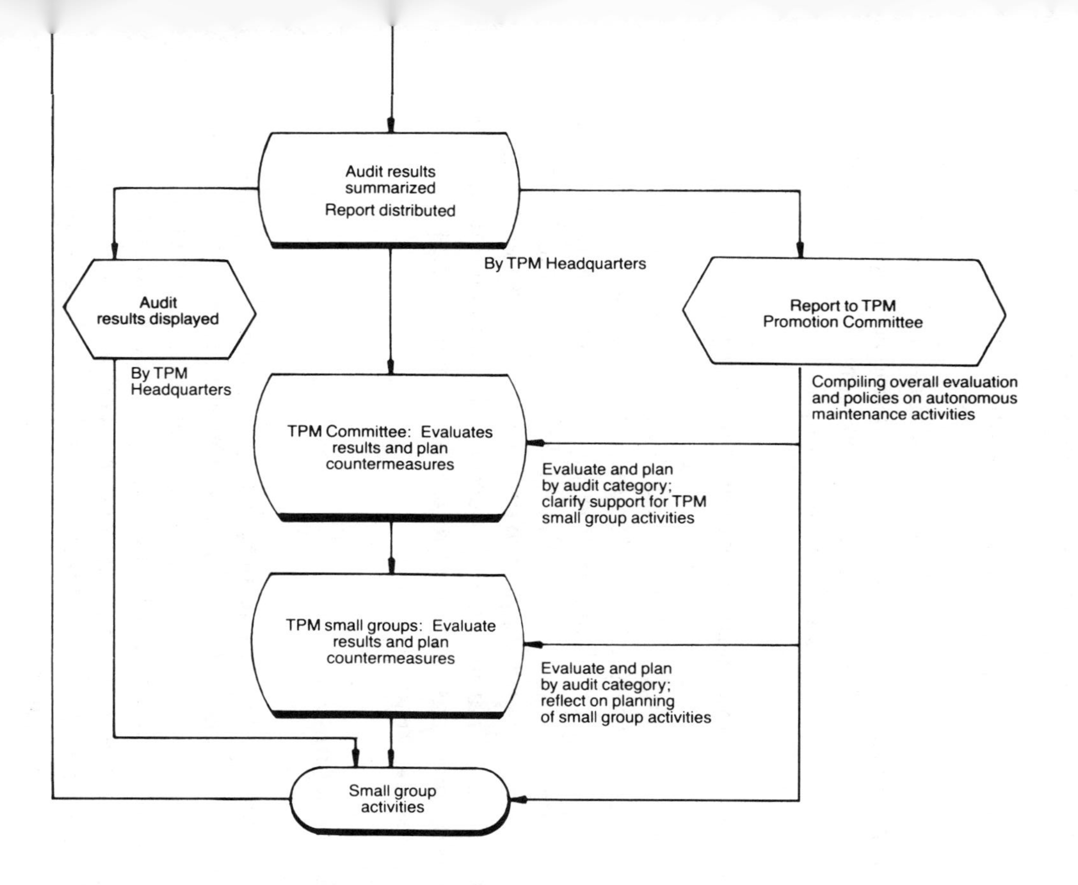

Figure 4-11. Autonomous Maintenance Audit Cycle

Always Aim for Concrete Results

Clear, appropriate, and concrete improvement themes and goals must be articulated at each stage to generate concrete results.

Operators Should Determine Standards to Follow

Operators must set their own standards and criteria for cleaning, lubrication, inspections, setup and adjustment, operation, and housekeeping. Moreover, they must acquire the skills needed to perform these tasks autonomously.

Management Should Audit Autonomous-Maintenance Progress

Managers and staff should audit autonomous-maintenance small group activity at every step and offer guidance and support in pinpointing problem areas. While managers are responsible for evaluating conditions in the workplace, they must also provide leadership and support to each TPM group. As each step in autonomous maintenance is successfully implemented, they should encourage operators to feel a sense of accomplishment (Figure 4-11; *see* pp. 214-215).

Use Model Projects

Select pieces of equipment or individual TPM groups to serve as models for the TPM development program as a whole. By allowing work in these groups (or on this equipment) to proceed one step ahead of other groups, managers and supervisors can anticipate important instructional and resource needs. The use of model projects will facilitate understanding at each stage of overall development.

Correct Equipment Problems Promptly

Most malfunctions that surface as a result of autonomous maintenance activities must be dealt with by the maintenance department. Treatment of these newly discovered problems and implementation of equipment improvement plans must be performed quickly, however.

Often, maintenance departments are swamped when they attempt to meet such large-scale demands in addition to conducting its own planned maintenance activities. If they cannot meet the new demands promptly, however, shop-floor conditions will not improve, autonomous maintenance will not progress, and the small groups will falter.

To avoid this potentially serious problem, the maintenance department must plan in advance to meet these increased demands effectively — by reevaluating their manpower allocations, scheduling, overtime, and the use of outside contractors.

Take Time to Perfect Autonomous Maintenance

It is vitally important to perfect each stage of autonomous maintenance before advancing to the next. If the activities involved in a particular step are rushed, some superficial progress will appear to have been made. The skills needed to maintain that progress will not be mastered, however; serious problems will surface later and the program may collapse.

REFERENCES

Gōtō, Fumio. "Promoting autonomous maintenance" (in Japanese). *Plant Engineer* 11-12 (June 1979 to December 1980).

5

Preventive Maintenance

Preventive maintenance is periodic inspection to detect conditions that might cause breakdowns, production stoppages, or detrimental loss of function combined with maintenance to eliminate, control, or reverse such conditions in their early stages. In other words, preventive maintenance is the rapid detection and treatment of equipment abnormalities before they cause defects or losses. It is preventive medicine for equipment.

Preventive maintenance consists of two basic activities: (1) periodic inspection and (2) planned restoration of deterioration based on the results of inspections. Daily routine maintenance to prevent deterioration is usually considered a part of preventive maintenance as well.

This chapter contains a discussion of intermediate and long-term planned maintenance activities conducted by the maintenance department: setting maintenance standards, preparing and executing maintenance plans, keeping maintenance records, and carrying out scheduled equipment restoration activities. It covers subsystems such as spare parts control, lubrication control, and control of maintenance budgets. The chapter concludes with a summary of equipment diagnostic techniques for predictive maintenance.

STANDARDIZATION OF MAINTENANCE ACTIVITIES

Maintenance activities ought to be standardized for several reasons:

1. The diverse maintenance activities — from routine maintenance and inspection to repair and maintainability improvement — cannot be performed effectively if individuals are left to carry them out in any way they like.
2. Maintenance techniques and skills take a long time to master. On the other hand, when only experienced workers are able to apply them, demands on the maintenance department exceed its capacity and maintenance goals cannot be achieved.
3. Maintenance work is generally less efficient than production work because it is essentially nonrepetitive and requires lengthy preparation and large margins for error. It relies heavily upon individual skills and is performed under difficult conditions. Individual workers must transport equipment and move around the factory frequently.

Standardization addresses each of these problems and is necessary for the consistent, efficient performance of maintenance activities. For these reasons, comprehensive maintenance standards and manuals incorporating a company's past experience and technology are indispensable. Such documents enable large numbers of workers, including new recruits, to do work that previously could be done only by experienced workers. This capacity to train and involve many individuals in maintenance work is the key to the development of a high-quality, efficient maintenance program.

Types of Standards

Equipment design standards, or simply *equipment standards*, are company standards for common equipment elements (*e.g.*, bearings, gears, valves, and flanges), standard methods of calculating equipment capacity, and so on.

Equipment performance standards, or *equipment specifications*, apply to equipment performance during operation. They indicate how equipment is to be operated and include its principal dimensions, capacity and performance, precision, functions, mechanisms, the materials its main parts are made of, the quantities of electric power, steam, and water needed for operation, and so on.

Equipment materials procurement standards cover the quality of equipment materials and parts. They are based on the equipment design standards and equipment performance standards. *Equipment materials inspection standards* provide standard testing and inspection methods for determining whether materials and parts used in equipment meet the standards.

Test run and acceptance standards indicate the acceptance and operating tests to be performed on equipment that has been newly installed, modified, or repaired.

Maintenance Standards

Equipment maintenance standards indicate methods for measuring equipment deterioration (inspection and test), arresting the progress of deterioration (daily routine maintenance), and restoring equipment (repair). There are separate standards for each maintenance function, including equipment inspection standards (inspection), servicing standards (daily routine maintenance), and repair standards (repair work).

Maintenance work procedures are the work procedures, methods, and times for inspection, servicing, repairs, and other types of maintenance work.

Equipment maintenance standards and maintenance work procedures are referred to collectively as *maintenance standards* (Table 5-1).

Equipment Maintenance Standards

As mentioned above, equipment maintenance standards include inspection, servicing, and repair standards.

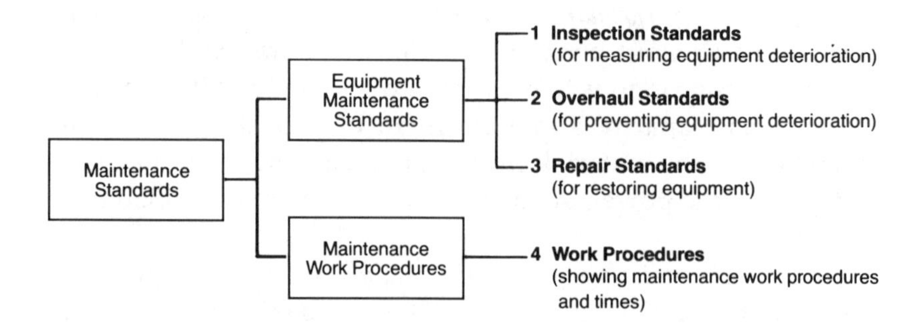

Table 5-1. Types of Maintenance Standards

Inspection standards. These are standards for inspecting equipment, in other words, techniques for measuring or otherwise determining the extent of deterioration. They specify the areas and items to be inspected, the inspection intervals, methods, measuring instruments, evaluation criteria, corrective action to take, and so on. They should include illustrative sketches and photographs where necessary.

Depending on the type of inspection, inspection standards can be classified by interval, as in routine inspection standards (for intervals of less than a month) and periodic inspection standards (for intervals of more than a month). They can also be classified by the item to be inspected (performance inspection standards and precision inspection standards) or by type of equipment (mechanical and electrical equipment inspection standards, piping inspection standards, and instrument inspection standards). Table 5-2 (*see* pp. 224-225) is an example of a performance inspection standard.

Servicing standards. These standards specify how servicing and routine maintenance done with hand tools will be carried out. They include methods and guidelines for different types of servicing, such as cleaning, lubrication, adjustment, and parts replacement. For example, lubrication standards will indicate the parts to be lubricated, methods of supplying lubricants, types and quantities of lubricants, lubrication intervals, and so on.

Repair standards. Repair standards specify conditions and methods of repair work. They may be drawn up separately for specific equipment or parts, or classified according to the type of repair work (*e.g.*, turning, finishing, piping, or electrical work). Repair standards usually include repair methods and available work hours and can be used as maintenance work standards.

Maintenance work standards. These standards are prepared for frequently performed work. They are helpful in measuring the efficiency of maintenance crew, estimating available work hours and reserve capacities, setting schedules, and training new workers.

Revision of Standards

Maintenance standards must be revised as equipment is updated and improved. Methods will change naturally as equipment is restored, revamped, or otherwise improved. Maintenance results should therefore be reviewed and the standards revised at least once a year.

Maintenance standards serve as a barometer to the technical level of the maintenance department. They should be continuously improved through practice. As better and better maintenance results are obtained, the data should be used to upgrade standards.

MAINTENANCE PLANNING

This section reviews the criteria for effective planned maintenance. Routine and periodic maintenance must be well planned and reasonable. In other words, it must be based on accurate assessments of equipment conditions and systematically mapped out, taking into consideration current and future priorities and resources and building in steps to ensure that the appropriate resources are available when needed. Cost-effective and efficient planned maintenance requires the close cooperation of all departments involved.

Types of Maintenance Plans

Maintenance plans are classified by period or project.

Performance Criteria for Centrifugal Pumps						
Item	Method	Measuring Instrument	Criteria			Interval
			Specified value	Control limit	After repair	
1 Vibration	A 1 B 1 b 3 3 a 2 C	Vibration indicator	10μ	40μ	15μ or less	1 × per 6 months
2 Bearing temperature	Measure temperature of bearing with surface thermometer during operation (3 + hours after startup)	Surface thermometer	Atmospheric temperature + 20°C	Atmospheric temperature + 40°C	Atmospheric temperature + 25°C or less	1 × per 6 months
3 Discharge pressure	Read indication and deviation on discharge-side pressure gage (fully opened discharge valve)	Discharge-side pressure gage	5 ± 0.5	4 ± 0.5	5 ± 0.53	1 × per 6 months
4 Motor current	Read indication and deviation on control panel ammeter (fully opened discharge valve)	Control panel ammeter	112	100	75	1 × per 6 months
5 Motor coil insulation	U V W IM Measure resistance between coil terminals and earth using 500-V megger insulation tester before dismantling motor	500-V megger	0.2 MΩ	0.4 MΩ	1.0 MΩ	1-2 × per year
		Remarks				

Table 5-2. Sample Performance Criteria for Industrial Machinery

Inspection Record									Date / Operating Time	
After repair	Inspected	After repair	Inspected		After repair	Inspected	After repair	Inspected	Classification	
									1	A
									2	
									3	
									1	B
									2	
									3	
									a	
									b	
									c	
									kg/cm²	
									A	
									U	
									V	
									W	
									Special items	
									Inspector	
									Signature	

Notes on preparation and execution of criteria

1. Specified values represent the rated capacity of the equipment
2. Control limits are set from considerations of product quality and repair costs. They cannot always be decided in advance and must then be determined from experience
3. The accuracy and performance after repair are those that can be restored economically. It is not always necessary to achieve the specified values
4. If inspection shows the values to be within the control limits, no repairs are needed. In this case, the "after repair" column on the inspection record can be left blank
5. This inspection standard can also be used as a repair standard during periodic repairs and as an equipment record

Annual maintenance plans should guarantee the reliability of equipment over its predicted lifetime, from installation to scrapping. Their preparation requires the coordination of production plans, subcontracting, and procurement of cost-effective spare parts. For this reason, inspection and maintenance plans should be drawn up from a long-range perspective.

Monthly maintenance plans are based on the annual maintenance plans and include improvement activity as well as specific actions to prevent breakdowns. Their purpose is to evenly allocate the required work among the available maintenance workers and firmly guide the work assignments and progress.

Weekly maintenance plans help manage the work of individual maintenance personnel.

Major maintenance project plans are individual plans for turnaround, large-scale revamping, or overhaul of specific equipment or plant areas. This type of plan includes procurement plans for maintenance spare parts, equipment, and other materials, as well as prudent arrangements for subcontracting, acquiring cost-effective spare parts or foreign-made equipment.

Equipment Maintenance Standards and Maintenance Planning

Effective maintenance planning is impossible without an accurate understanding of equipment conditions. Obtaining a comprehensive view of actual equipment conditions is difficult, but it is helpful to follow the equipment maintenance standards in conducting inspection. Both daily routine inspection and periodic inspection will contribute to this effort. Routine inspection uses the senses (mainly looking, listening, and touching) to detect abnormalities and prevent breakdowns before they can take place. It is usually carried out by operators while the equipment is in operation, as part of their autonomous maintenance activities.

By contrast, periodic inspection is usually carried out by maintenance workers while equipment is shut down. Using various measuring instruments, they measure equipment deterioration, maintain precision, and replace parts before failure.

The intervals for overhauls, parts replacement, and so on, are determined on the basis of this inspection data. Maintenance plans can then be prepared that specify when and how periodic servicing and repairs are to be executed.

Inspecting every item of equipment and replacing parts indiscriminately will not result in zero breakdowns, however. On the contrary, the cost of production losses and parts replacement would be far greater than the amount saved. Better and more economical maintenance results are achieved by concentrating on the most important items of equipment. Collect data from the past one or two years, estimate future production and equipment plans, and draw up maintenance plans starting with the equipment items that will yield the best results. Table 5-3 is an example of an equipment priority ranking chart.

Bear in mind that priorities, once decided, will not necessarily remain the same. They may change in response to altered production plans, new equipment installation or revamping, and the results of maintenance improvement activities. Therefore, plan to revise priorities every one to two years.

Preparing Annual Maintenance Plans

Annual maintenance plans are designed to ensure equipment reliability over the long term. They should not be limited to one year but should incorporate servicing items spanning two to three years (Table 5-4).

To prepare annual maintenance plans:

1. Determine What Work Is Required

The most important task in preparing maintenance plans is identifying all the work that must be done over the year. This list must be revised annually. Required work may include

- *statutory regulations:* work required for safety, pollution control, and so on.

Area	Item	Evaluation				Evaluation Standard
Production	1. How often is the equipment used?		4	2	1	80% or above: 4 59% or below: 1
	2. Is there backup equipment?	5	4	2	1	No (or) Yes, but it takes too many manhours: 5 • available at other plants: 4 • covered by stock: 2 • backup equipment exists: 1
	3. How high is the dedication? (the proportion of products of a similar type produced by the equipment)		4	2	1	100–75%: 4 0–35%: 1 35–75%: 2
	4. To what extent will a failure effect other processes?	5	4	2	1	Affects the entire plant: 5 Strongly effects other processes: 4 Only effects this machining center: 1
Quality	5. Value of monthly scrap losses (burned rubber, wasted cloth, wasted production)		4	2	1	(burnt rubber) over $1000, $500–1000, under $500 (wasted cloth) over $5000, $2000–5000, under $2000 (wasted production) over $1000, $500-1000, under $500
	6. How will the process run on this equipment affect the quality of the finished product?	5	4	2	1	Decisively: 5,4 Not significantly: 1 Somewhat: 2
Maintenance	7. Frequency of failures in terms of cost of monthly repairs?		4	2	1	over $5000: 4 Under $3000: 1 $3000–5000: 2
	8. Mean time to repair (MTTR)		4	2	1	MTTR over 3 hours: 4 Under 1 hour: 1 1–3 hours: 2
Safety	9. To what extent does a failure affect the work environment? (noise, etc.)	5	4	2	1	Can be life-threatening: 5 No significant effect: 1 Stops work: 4

A (priority-ranked equipment) (30 or more points) B (20–29 points) C (19 points or less)

Table 5-3. TPM Priority Management Table for Rating Equipment

Type of maintenance:
- ■ Turnaround day
- ○ Planned weekday shutdown maintenance
- △ Detailed inspection and adjustment

Degree of difficulty:
A = Designated technician/Work by designated vendor
B = Work mainly by facility section
C = Work by maintenance worker (PM) or operator (OP)

Equipt:
Rolling unit

Dept:
Hiratsuka tire plant

Equipment and parts names	Maintenance work	Diff.	MTBF	MTTR	Dept. Responsible (In-house/outside)						Yr/mo	1980												Remarks
					Mech.	Elec.	Piping	Instr.	Lubr.	PM	OP resp.	1	2	3	4	5	6	7	8	9	10	11	12	
Roller press P/R	Abnormal noise, load current	B	6 months								○				○						○			
〃	Replace speed reducer oil	C	6 months								○						○						○	
Fastener	Pressure regulator	B	6 months					○								○						○		
〃	Check cylinder gaskets	C	1 year								○								○					
EPC unit	Actuation check	B	6 months				○							○						○				
Expander unit	Actuation check	C	6 months								○			○						○				
Compensator	Actuation check, load measurement	B	1 year					○							○						○			
Preheat D/M	Bearings	B	6 months						○				○							○				
〃	Speed reducer oil check and sampling	B	6 months						○								○						○	
〃	Motor load current	B	3 months			○								○			○			○			○	

Tsukui, "Annual Maintenance Calendar" (in Japanese), *Plant Engineer* 12 (June 1980)

Table 5-4. Sample Annual Maintenance Calendar

- *equipment maintenance standards:* work determined by precision control requirements and the results of deterioration measurement
- *breakdown records:* maintenance work to prevent recurrence of breakdowns
- *previous year's annual plan:* outstanding work due to schedule changes
- *work orders received from shop floor:* work deemed necessary based on abnormalities records

2. Select Work to Be Done

Rank work in the order of its importance and establish priorities. Focus on the most important items.

3. Tentatively Estimate Maintenance Intervals

Make trial estimates of the life spans of all equipment, component by component and part by part, and decide the maintenance intervals (the TBO, or time between overhauls), preferably using breakdown records (MTBF analysis charts, etc.).

4. Estimate Work Schedules and Maintenance Times and Costs

Use the annual production plans and equipment performance targets to estimate the number of shutdown days and the time required for maintenance work, and confirm these figures against the budget.

5. Check Procurement and Work Arrangements

Confirm the arrangements for materials and hard-to-get spare parts and for work to be done by outside manufacturers and contractors. Determine whether specially qualified personnel are needed.

Preparing Monthly Maintenance Plans

Monthly maintenance plans are action plans for carrying out the work required by the annual maintenance plans.

Prioritize work. For the best results, rank the work in the following order of priority:

1. monthly work indicated by the annual maintenance plan
2. work indicated by an analysis of breakdown and inspection records
3. work indicated by daily inspection and improvement requests from the production department
4. layout changes and installation plans for jigs and tools
5. plans for improving product quality and safety

Other work should also be included in the plan, such as work orders from various departments, salvaging reusable parts, preparing for maintenance work, and cleaning up afterward. These needs should be addressed at monthly PM meetings. *At meetings, each department must be willing to take on whatever work it can do.*

Estimate labor and costs. Once the details of the work have been determined, estimate the manhours and costs. If the maintenance work is organized by occupational category, estimate the manhours separately for each occupation, calculate the totals, and adjust the workload accordingly.

Balance workloads and prepare schedules. To level out the workload over the month, split the work into weekly units, starting with the work to be done on designated days or during planned shutdowns. Apportion the remaining work by its estimated duration and by other considerations. Table 5-5 is an example of a monthly maintenance plan.

Weekly plans should spell out the work assignments for individual maintenance staff. To simplify progress control, they should be designed so that work can be checked off as it is completed.

No.	Dept.	Equipment	M/c (shutdown time)	Type of work						Details of work
				M	E	P	I	PM	Other	
1 2	Rolling unit	Z calendar roller	8 – 16h 〃	○ ○	○	○				1 Change strip field at No. 1, 2 roller side 2 Change position of fixed roller for winding replacement
3	〃	〃	〃	○	○	○				1 Install air bleed device at No. 3 roller (see drawing) 2
4 5	〃	〃	〃	○	○	○ ○			Ⓢ	1 Install monorail for replacement of calender roller beam, plus related work 2 Check oil cooler of main body crossing-roller hydraulic unit
6 7	〃	〃	〃		○					1 Check insulations of high voltage, booster MG and D motors 2 Perform sampling and calibration of B-ray thickness gage (annual plan)
8 20	〃	〃 Z calendar roller	Continuous			○		○		1 Check performance of heat-exchanger shell side (check safety valve) (annual plan) 2 Make endless seams in No. 1, 2, 3, 4 roller belts
9	〃	D-3 roller	8 – 16h	○	○					1 Improve cutter 2
10 11	〃	Sheet calendar (no. 4)	〃	○ ○	○					1 Replace feed calender drive motor 2 Improve drive unit for top roller adjustment (north side)
12	〃	〃	〃						Ⓢ	1 Operator-side monorail (cost estimation and vendor negotiation) 2
13	〃	Hoists	〃	○					○	1 Periodic inspection 1 × /6 months–11 units 2
14	〃	Z calendar roller	〃	Ⓛ						1 Flush out calender roller metal circulating tank and reuse oil 2
15	〃	D-4 roller	〃	Ⓛ						1 Replace oil in speed reducer gears as per sampling result

Tsukui, "Annual Maintenance Calendar," *Plant Engineer* (June 1980)

Table 5-5. Sample Monthly Maintenance Plan

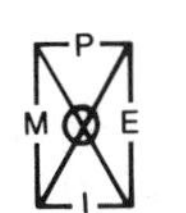

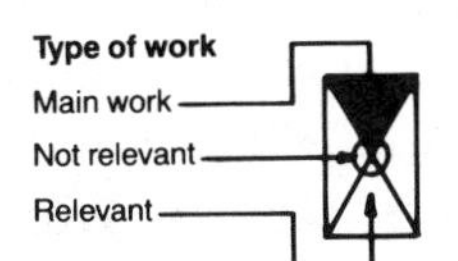

○ = Includes in-house fabrication
Ⓐ = 4 groups — 3 shifts
⊙ = 2 groups required for in-house fabrication
Ⓢ = Subcontracted work
Ⓜ = Machine shop
Ⓛ = Lubrication
PP = previous process

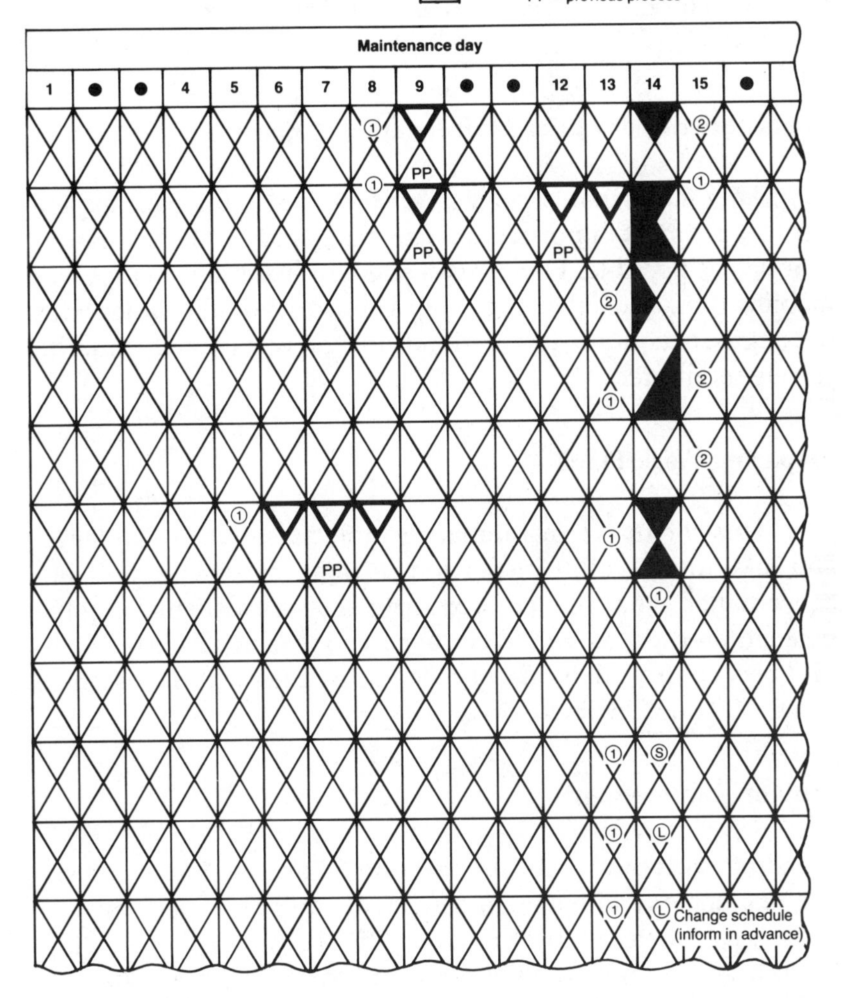

Planning for Major Maintenance Projects

Major maintenance projects are large jobs, such as periodic repairs or turnaround work, that require equipment to be shut down for an extended period. Since this kind of project often means large production losses, separate work plans and progress control methods are needed to ensure that the project proceeds efficiently.

Characteristics of Major Maintenance Projects

Major maintenance projects are time-consuming and expensive, so the most important objective is to reduce their duration as much as possible. Plans are needed for materials, manpower mobilization, and subcontracting. The work volume varies with the degree of equipment deterioration, and work quality must be closely monitored.

Preparing Work Plans

Plans for major maintenance projects should be detailed and take into account inspection results, revamping, and improvements. Since the work consists of a series of smaller jobs, detailed procedures should be prepared for each job. To facilitate progress control, draw up detailed schedules with milestones, using PERT or CPM networks.

Project Management

Projects typically consist of periodic overhaul and preventive maintenance to maintain consistent equipment performance as well as construction of new facilities, expansion of existing facilities, revamping, replacement, and so on. *Project management* means performing the work in a planned and cost-effective fashion. For better project management, focus on these four key points in the execution of monthly maintenance plans and major maintenance project plans:

1. Identify Problems

To ensure that planned work proceeds on schedule and without waste, it is important to anticipate problems that may arise and clarify any limitations under which the work must be done. The following types of limitations may need to be considered:

- equipment shutdown schedules and designated maintenance days determined by production plans
- available maintenance manpower and capacity
- maintenance cost targets and budgets
- capacity by skill level (number of experienced and inexperienced workers)
- whether work is to be done on a work day or over holidays
- whether the work is subcontracted or done in-house
- whether critical spare parts can be delivered and inspected when needed

2. Confirm Administrative Details

The most common reasons monthly plans fall behind schedule are administrative: forgetting to order, delays in ordering, insufficient expediting of orders, and so on. Time is also lost when checking specifications of parts after they have been delivered and deciding where to store them has not been arranged in advance.

Confirm subcontracted work well in advance and take care in arranging for maintenance equipment, jigs, and tools. Contact everyone involved when changes in the plan require cancellation or postponement of work and coordinate with the production control department when work requires equipment shutdown.

3. Implementing Major Maintenance Projects

Major maintenance projects require the cooperation of a number of different departments (*e.g.*, maintenance, production engineering, purchasing, subcontractors). Appoint representatives from each department to monitor the project progress and identify problems, and hold coordination meetings to discuss corrective action. Keep an eye on inspection results and modify or change the project plan as the occasion demands.

4. Progress Control

The maintenance supervisor's most important job is to ensure that the work proceeds according to schedule. The following key points of progress control should be observed:

- Estimate the manhours required and cumulate individual workloads.
- Identify the difference between estimates and actual manhours and use this data to increase the accuracy of future estimates.
- Confirm and follow up administrative arrangements.
- Assign work according to the individual skills of the workers.
- When workers are in groups of two or three, mix the skill levels so that they can learn from each other.
- To keep workers from standing idle if breakdowns occur, plan improvement work or work that can be done at any time.
- Analyze other requirements for keeping the work on schedule and take appropriate action.

Increase Maintenance Efficiency by Improving Maintainability

Improving equipment maintainability increases the efficiency of maintenance work and reduces repair times. Consider the following strategies for increasing maintainability:

- division into appropriate subassemblies for ease in dismantling and reinstalling
- prefabrication
- speedy and accurate communication
- improved transportation and material-handling equipment

- standardized parts and improved jigs and tools
- maintenance intervals that are balanced by scheduling related projects together
- worker awareness of the key factors in quality of maintenance

KEEPING AND USING MAINTENANCE RECORDS

Documenting the results of maintenance is one of the most important maintenance activities. Routine maintenance work is so varied, however, that keeping records on each task would be extremely difficult. Such comprehensive documentation may not be necessary.

Why Keep Maintenance Records?

The quality of a factory's maintenance is revealed by its maintenance records. Some factories maintain and use many kinds of organized maintenance records; others lack proper routine maintenance reports and make little use of those they do have.

The format for maintenance records is not fixed; the types and contents of maintenance records can be arranged to dovetail with a particular plant's management standards. However, everyone must understand the purpose of keeping records — why they are kept, what is being controlled, and how they should be used.

The Flow of Maintenance Records

Always feed back the results of maintenance activities quickly into the records system so the results can be incorporated in subsequent maintenance plans. Maintenance quality and performance levels can be continually raised by repeating the Plan-Do-Check-Action (PDCA) management cycle.

Figure 5-1 shows how maintenance records can be classified according to the types of maintenance activities and the stages of the management cycle.

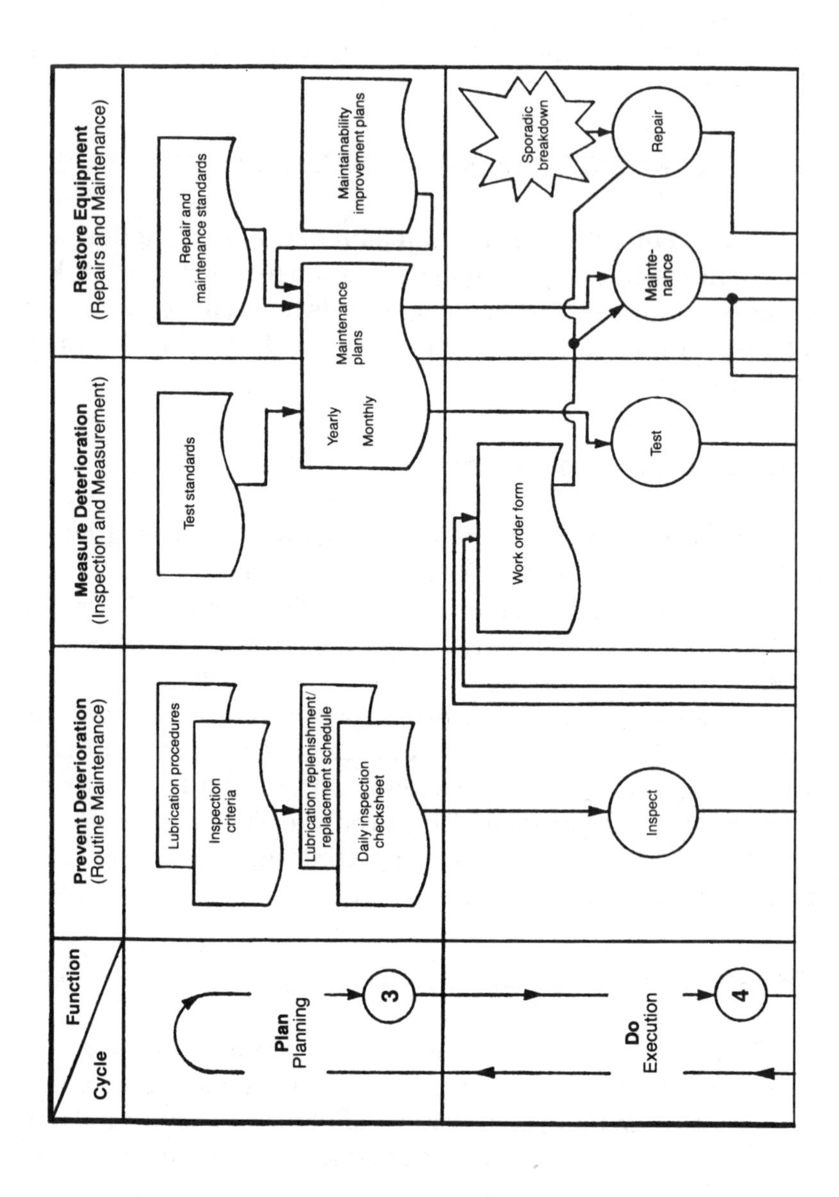
Function
Cycle
Prevent Deterioration
(Routine Maintenance)
Measure Deterioration
(Inspection and Measurement)
Restore Equipment
(Repairs and Maintenance)
Plan
Planning
3
Lubrication procedures
Inspection criteria
Lubrication replenishment/ replacement schedule
Daily inspection checksheet
Test standards
Yearly
Monthly
Maintenance plans
Repair and maintenance standards
Maintainability improvement plans
Do
Execution
4
Work order form
Sporadic breakdown
Inspect
Test
Mainte-
nance
Repair

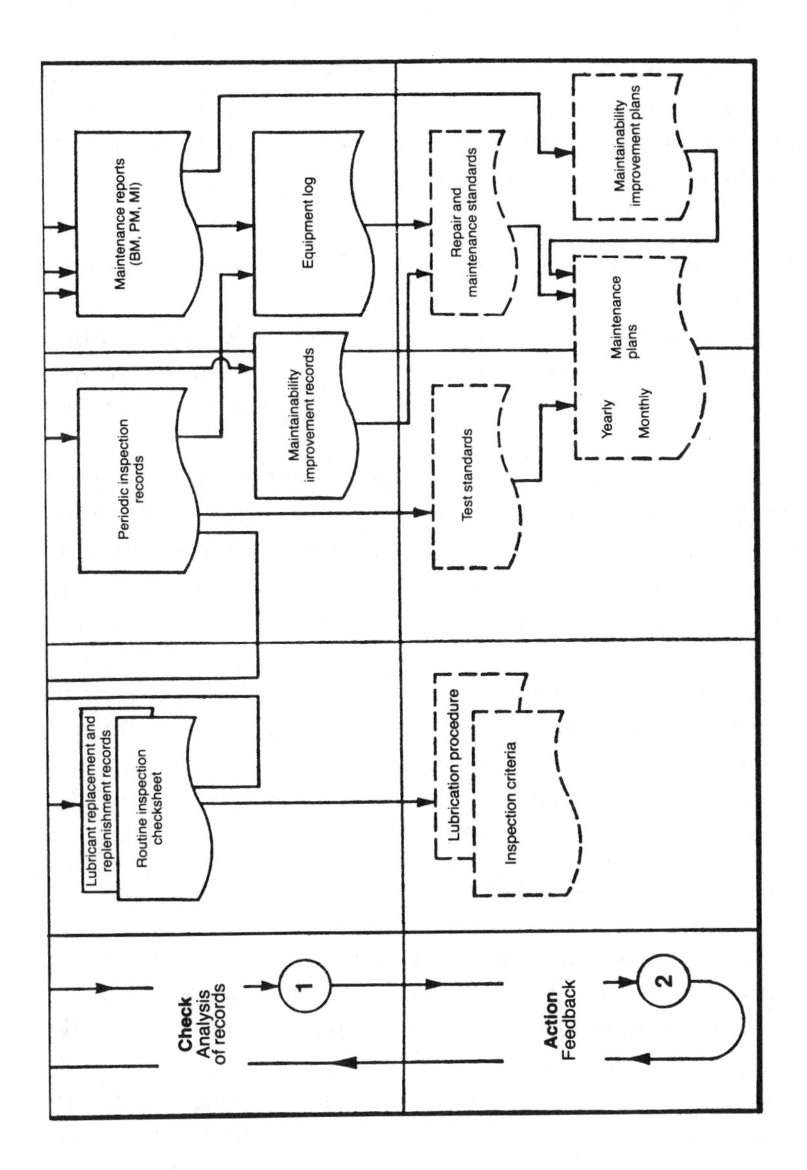

Figure 5-1. Maintenance Recording/Execution Flowchart

Types of Maintenance Records

How maintenance activities are carried out depends on plant policies, managerial levels, and so on. Thus, the variety in types of maintenance records prevents generalization about which are needed in particular cases. The following discussion outlines the basic maintenance functions and the types of records minimally required in the practice of TPM.

Routine Maintenance Records

Operators record routine maintenance aimed at preventing equipment deterioration and maintaining basic equipment conditions in the departments where the equipment is used. For example:

Routine inspection records are checked off daily, weekly, or monthly according to the equipment, part, or item to be inspected as required by the daily inspection standards. If an abnormality is discovered in the course of an inspection, it is dealt with according to a specified procedure.

Lubricant replenishment and replacement records are required, because intervals between replenishing and replacing lubricants vary from part to part. It is easy to overlook some lubricating points, especially when there are a lot of them. Therefore, it is best to keep daily lubrication records. (*See* Tables 5-6 and 5-7.)

Periodic Inspection Records

The results of equipment deterioration measurements are typically recorded by designated maintenance department personnel. In periodic inspections (statutory inspections, disassembly inspection, precision checks, etc.), criteria are needed to indicate how much wear is allowable before repairs must be made. The inspection standards lay down the dimensional tolerances, usable maxima and minima, inspection methods, and other conditions. Such inspections are usually carried out as part of annual maintenance schedules. The measurement data obtained form the periodic inspection records.

Plant: ____________________ Maintenance Section: ____________________

Equipment or device name, pump specification, instrumental specification, etc.	Drawing No.	Lubricant name	Capacity (l) × no. of reservoirs	Lubrication method	Part to be lubricated, number of lubrication points, type and number of distribution valves, number of cylinders, etc.	Lubricant replacement and replenishment standards				Periodic test, interval
						Replace lubricant (recycling methods)		Replenish lubricant (once-through methods)		
						Quantity (l)	Interval	Quantity (l)	Interval	

Table 5-6. Lubrication Control List

Equipment Specification

Plant/workshop

<table>
<tr><td colspan="2">Name of equipment:</td><td colspan="2">Type of lubricant:</td><td colspan="2">Reservoir capacity: l</td><td colspan="3">Amt. of lubricant in pipes: l</td></tr>
<tr><td>Pump</td><td>Model:</td><td>Mfr.:</td><td>No.:</td><td colspan="2">Pressure: kg/cm²</td><td>Motor</td><td>Wattage: kW</td><td>Voltage: V</td></tr>
</table>

Inspection/Malfunction Record

Date	Quantity of lubricant supplied	Describe malfunction	Describe action taken

Procedure for completing form:
1. This form must be prepared for reservoirs of 100 l or more (hydraulic equipment) or for 500 l reservoirs (other equipment)
2. Be sure to record quantity of lubricant supplied, whether or not malfunctions are found
3. Briefly record details of malfunctions and action taken

Onuma, "Lubrication Control" (in Japanese), *Plant Engineer* (November 1981).

Table 5-7. Lubrication Control Sheet

Maintenance Reports

These record repairs and servicing performed to restore equipment to its original condition. They are usually kept by the responsible maintenance department personnel. At the intervals specified in the annual maintenance plan, the work is performed and recorded in accordance with repair and maintenance standards. Breakdown repairs, the only type of work that must be performed immediately, are also recorded on maintenance reports.

Although the duties of the maintenance department include planned maintenance, maintainability improvement, breakdown repairs, and so on, a single report format should be used for each type of work (Table 5-8; *see* pp. 244-245). Occasionally, serious breakdowns may be reported separately to the relevant departments on an accident report form, but this should be done only when necessary.

Maintainability Improvement Records

Maintainability improvement modifies equipment to increase its reliability or maintainability. This activity is furthered by collecting and analyzing breakdown data from maintenance reports, implementing improvement proposals submitted by line workers, identifying equipment that breaks down frequently, and considering ways to prevent recurrence.

The results of maintainability improvement activity should always be compared with data obtained before improvement to see whether or not the anticipated effects were achieved. For this reason, it is more convenient to keep maintainability improvement records on separate forms.

MTBF Analysis Records

The purpose of equipment improvement is to reduce the maintenance work required and to increase its efficiency. To promote this, record each type of work performed on a particular piece of equipment on a separate card and organize the information on an MTBF analysis chart (Table 5-9).

MTBF analysis charts help to clarify and classify the occurrence of breakdowns. They show at a glance the breakdown frequency of each machine and part.

①	②
③	
④	⑤
⑥	⑦

* Fill out maintenance work card for every job.

① Date maintenance performed
② Work done by (name)
③ Name of equipment and part
④ Breakdown details
⑤ Describe action taken (maintenance details)
⑥ Manhours (personnel × min.)
⑦ Equipment downtime (min.)

Equipment MTBF analysis chart

Analysis duration:
Start: _______ End: _______

Name of part	Maintenance work performed					
A–a	☐	☐				
A–b	☐	☐	☐	☐	☐	
A–c	☐	☐	☐			
B–a	☐					

* Post maintenance work cards on MTBF analysis chart.

Table 5-9. MTBF Analysis Chart

Equipment Logs

Equipment logs are maintained over the lifetime of the equipment and are the equivalent of a doctor's medical charts. They offer an up-to-date history of a piece of equipment that begins with its dates of purchase and installation and records all accidents and major repairs from startup to the present. A complete log will include the dates, location, details, and costs of all major breakdown repairs, periodic maintenance, and maintainability improvements, as well as the names, models, sizes, numbers,

Daily Inspection Report

Date:

Equipt.	**A** Malfunction	Predicted downtime (DT)	Action				Equipt.	**B** Adjustment/Minor Repair
			P	**Q**	**R**	**S**		

Equipt.	**C** Accident	Type	Recorded	Work time	Manhours for repair	Downtime

Table 5-8. Daily Maintenance Record

Sect. Mgr.: ____________	Sup.: ____________
Fore.: ____________	Inspector: ____________

Class.	Predicted DT	Remarks	Action		
				Details	Symbol
			P	Repair pending until to next shift Repairs pending until holiday work	◎ ○
			Q	Inspection pending	○
			R	Holiday work (other section) Holiday work (section in charge)	◎ ○
			S	Spare parts ordered	○
			Remarks		

Parts used	Remarks	Records and monthly report		Types	
		Monthly report	V	Inspection error	A
		Down-time reports		Lubrication error	B
		Major work reports		Replacement error	C
		Accident reports, etc.		Repair error	D
		Maintenance records	W	Installation error	E
		Machine history	X	Acceptance error	F
		Work control list	X_1	Design error or fault	G
		Revisions and additions to standards	Y	Deterioration	H
				Managerial error	I
		Improvement suggestions	Z	Unforeseeable event	J
				Minor repair	K
				Spare parts error	L
				Operator error	M
				Other group's error	N
				Other	O

Fukunaga, "Introduction to PM," *Plant Engineer* 11 (August 1979)

and manufacturers of spare parts. Individual equipment logs should be kept in a form that will last until the equipment is scrapped. (*See* Table 5-10.)

Lost performance can be restored by repair work, but this costs money. When a machine has deteriorated past the limit of its economic life, it must be replaced. Thus, an important function of the equipment log is to help set standards for equipment replacement.

Maintenance Costs Records

Maintenance costs include labor, materials, subcontracting costs, and so on. To control maintenance budgets, each maintenance expenditure must be recorded by item and by usage as it is made. Since these records help control the annual maintenance budget, a running total of the amount of budget consumed is also entered as each item of work is completed. Maintenance activities are carried out according to repair orders (maintenance requests or work orders), and the costs are totaled after passing through the company's accounting system.

Using Maintenance Records

The types and contents of maintenance records have been discussed above. Their uses are summarized in Table 5-11.

Record-Keeping Precautions

The following precautions should be taken when keeping records:

Clarify the *5Ws* and *1H*

Sometimes maintenance records are not used effectively because they are poorly written and hard to understand. It is vital that the forms answer the *5W*'s and *1H* (who, what, when, where, why, and how). The records must show clearly *who* is to

Asset No.	
Equipt. name	
Model	
Spec.	
Main motor	

Manufacturer	
Date of manufacture	
Vendor	
Date of purchase	
Purchase price	

Place of Installation	
Date	Section/line name

Date	Periodic maintenance/Improvement				Cost
	In-house/ outside	Part	Details	In-house manhours used	

Date	Major breakdown repairs				Cost
	In-house/ outside	Part	Details	In-house manhours used	

Table 5-10. Equipment Log

	Function	Type of Record	Contents
1	Prevent equipment deterioration	Daily inspection checksheet	Daily record of presence or absence of abnormalities (visual inspection of equipment during operation)
2		Lubrication record	Record replenishment of lubricants and replacement of contaminated lubricants
3	Measure equipment deterioration	Periodic inspection record	Record of measured degree of deterioration and wear Analyze as necessary
4	Restore equipment	Maintenance report	Details of repair of sporadic breakdowns, planned maintenance, and maintainability improvement
5		Maintainability improvement record	Record of maintainability improvement plans, execution, and results
6		MTBF analysis chart	Record of all types of maintenance work, *e.g.*, repair of sporadic breakdowns, replenishment and replacement of lubricants, periodic maintenance, etc.
7	Document equipment lifetimes	Equipment log	Details and cost records for major breakdown repairs, periodic maintenance, and maintainability improvement
8	Control maintenance budget	Maintenance cost record	Breakdown of maintenance labor costs, materials costs and subcontracting costs Cost breakdown for each piece of equipment

Table 5-11. Maintenance Records and Their Uses

Use	Personnel Responsible	Remarks
Deal with abnormalities and report to superiors and maintenance department	Line operator	Can also be used for lubrication records
Improve lubricating methods and check lubricant consumption	Line operator	
Carry out repairs and maintenance if measurements show that control limits have been reached	Designated maintenance personnel	Control limits are specified in inspection standards
Obtain breakdown statistics and decide priorities for maintenance work Infer causes of breakdowns and take measures to prevent their recurrence	Maintenance personnel responsible	
Promote standardization of improved procedures and revise original drawings Use as improvement case study material	Maintenance or engineering personnel or staff	Deal with similar items of equipment together
Extending maintenance intervals and improving efficiency of repair work	Line operators, maintenance department, resident subcontractors, etc.	
Provide cost data on which to make decisions about equipment replacement and investment based on life cycle costs	Maintenance department personnel or staff	
Control maintenance budget, identifying priorities for reducing costs, and planning countermeasures	Maintenance, materials and purchasing department personnel and staff	Data forwarded to accounting dept.

fill them out (maintenance staff or operator), *what* should be recorded (description or cause of breakdown), *when* they should be filled out (daily or as necessary), *where* they should be filled out (in the maintenance department or on the production line), *why* they should be filled out (what they will be used for), and *how* they should be filled out (using words, numbers, or sketches). If the answers to any of these questions are unclear, it will be impossible to use the records effectively.

Record Breakdowns

Describe conditions at breakdown. Describe the condition of the relevant part in the equipment where the breakdown occurred (*e.g.*, breakage, deformation, breaking off). Because the causes of a breakdown are inferred from the visible signs of the breakdown and are not always easily identified, treat the breakdown and its causes separately and describe the breakdown in as much detail as possible.

Identify abnormal conditions leading up to breakdown. Some breakdowns appear suddenly and without warning, but most are preceded by warning symptoms such as unusual noise, excessive vibration, or overheating. Identifying these abnormal conditions makes it easier to infer the causes of the breakdown and helps in planning action to prevent recurrence.

Illustrate with diagrams or sketches. The location of a breakdown may be difficult to describe in words. Indicating the location and description of the breakdown through sketches or diagrams makes the report easier to understand. Another good method is to photocopy the original equipment drawings and indicate the location and nature of the breakdown on the copies. This is easy to do and is easily understood by anyone seeing the drawings.

Computerization of Maintenance Records

The collation and analysis of maintenance records takes many administrative manhours, but summary reports cannot be used effectively if they are delayed. Computerization permits the rapid analysis of large numbers of maintenance records and can provide the appropriate information when it is needed.

The first step in computerizing records is to simplify and standardize the current administrative procedures. Thereafter, proper use of computers reduces administrative manhours and makes needed data more accessible. It is also aids in planning appropriate action.

SPARE PARTS CONTROL

Everyone has a favorite horror story about large production losses caused when a warehouse full of unnecessary materials and spares lacked one essential part. Unfortunately, breakdown analysis often overlooks the poor management of parts control.

Purpose of Spare Parts Control

In reviewing the factory's spare parts control, ask why spare parts must be stocked and why control is necessary; ask whether the following three purposes of parts control are being fulfilled:

- Promote increased equipment reliability and extend equipment lifetimes through the purchase, fabrication, and storage of spare parts.
- Ensure that necessary spare parts are available whenever needed and thus minimize planned maintenance downtime and production stoppages due to breakdowns.
- Reduce inventories, ordering and acceptance costs, and storage costs.

Establish a parts control system suited to the production characteristics of the company that fulfills these purposes and review it periodically.

Classification of Maintenance Materials

Classification is the first step in management and control. First, identify the current status of the inventory and begin to classify it. The classification of maintenance materials is shown in Figure 5-2. *Operating materials* include measuring tools and consumables, for example. The production department is responsible for stocking, storing, and controlling these materials.

Maintenance materials can include necessary and unnecessary materials. Unnecessary materials should be scrapped and disposed of quickly (*e.g.*, broken parts, old equipment that is no longer useful or usable, and replaced parts that are no longer needed). Necessary parts come under maintenance store control or central store control, depending on where it is most convenient for easy control.

Maintenance stores are best located near the production line for the convenience of maintenance activities. Materials to be stored in maintenance stores include

- spare equipment for regular replacement
- spare parts for sudden breakdowns
- reserves always kept in stock
- reusable parts salvaged by the maintenance department
- maintenance tools

Spare equipment may include pumps, motors, speed reducers, and so on. The maintenance department should service and keep historical records on these items. Spare parts should be classified as priority parts or common parts. Priority parts are the most important parts of the most important equipment — in other words, parts whose unavailability would cause serious production stoppages. Very expensive parts also fall into this category. The maintenance department's parts-control system must place the greatest emphasis on priority parts.

Common parts such as bolts, nuts, and other low-cost parts should be supplied automatically. Tools should be kept in a toolroom and loaned out. They should be restocked by ordering from the tools section.

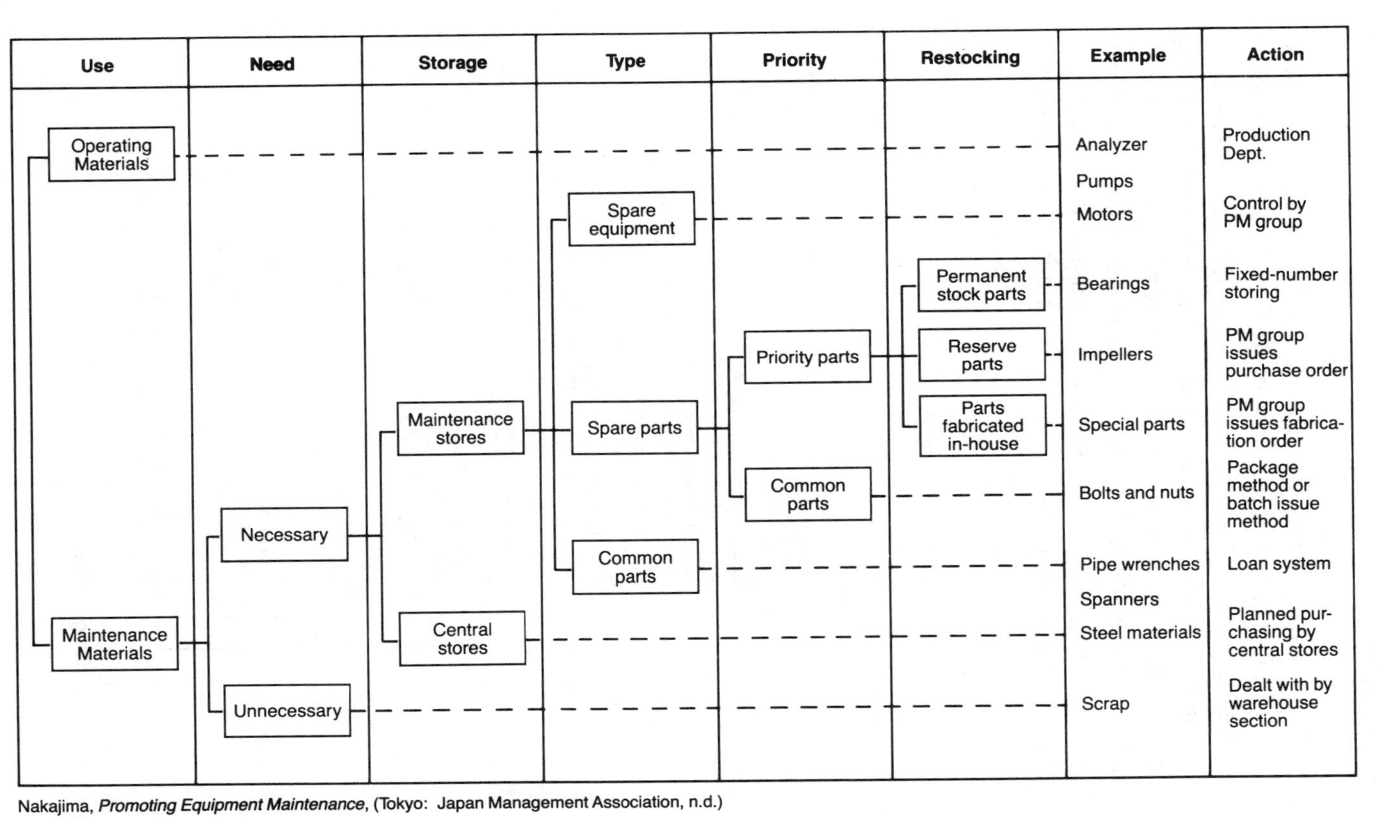

Nakajima, *Promoting Equipment Maintenance*, (Tokyo: Japan Management Association, n.d.)

Figure 5-2. Classification of Maintenance Materials

Ordering Methods

Methods for ordering spare parts can be broadly classified into individual orders and permanent stock methods. Individual orders are for parts that are only ordered as they are needed.

Permanent stock is material kept on hand continuously, and permanent stock methods restock automatically whenever designated quantities for such parts drop below a certain inventory level. Permanent stock methods are either based on a fixed quantity or a fixed period, depending on the method of restocking.

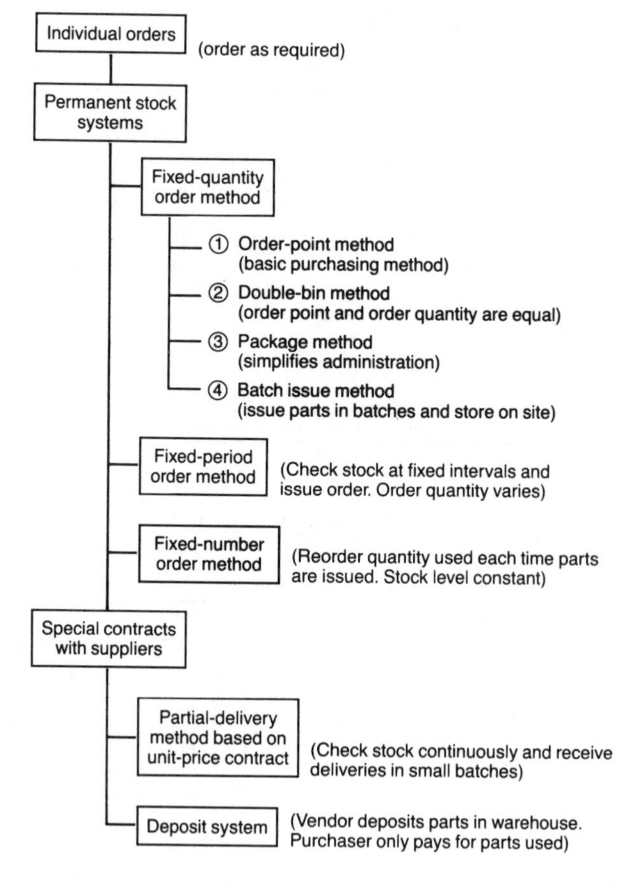

Figure 5-3. Restocking Methods

Other new methods of replenishing permanent stock include the partial delivery and deposit systems, which are arranged through special contracts with suppliers. These new systems simplify ordering, delivery, storage, and administrative procedures. Figure 5-3 summarizes the various ordering methods.

Permanent Stock Methods

Figure 5-4 shows the basic types of permanent stock methods: how inventory levels vary, how much to order at what inventory level, how long delivery takes, and how much spare stock is left for different rates of consumption.

The different methods are explained below:

Order-point method. This basic fixed-quantity method is the most common method of ordering. It is suitable for small spare parts with a fairly stable rate of consumption. A fixed quantity of stock is ordered when the inventory has dropped to a preset level (the order point P), with the aim of taking delivery when the minimum stock level (m) is reached.

Double-bin method. In this method, also called the double-box or double-shelf method, the order point and the quantity ordered are the same. Two containers, each holding the order quantity (the order point), are prepared for each part, and an order is issued when one container becomes empty. This method is suitable for bolts, nuts, and other low-cost parts stocked in large quantities.

Package method. In this method, parts are kept in packages (paper packages, boxes, or bundles) holding a quantity equivalent to the order point. The loose parts are used first, and an order is issued as soon as a new package is opened, using the order form attached to the package.

Since no record of parts issued is kept in either the double-bin or the package method, administration is simplified. Both methods are limited to low-cost items, however, since they do not permit a continuous check on the number of parts left in stock.

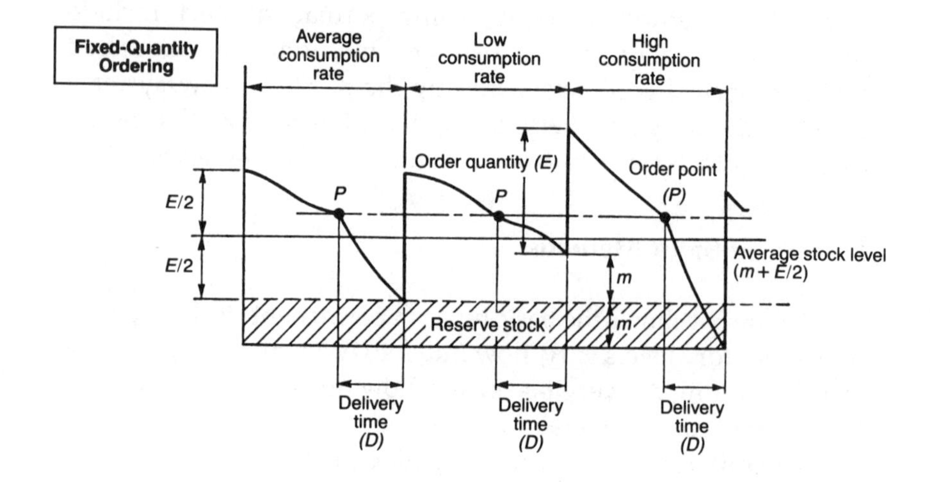

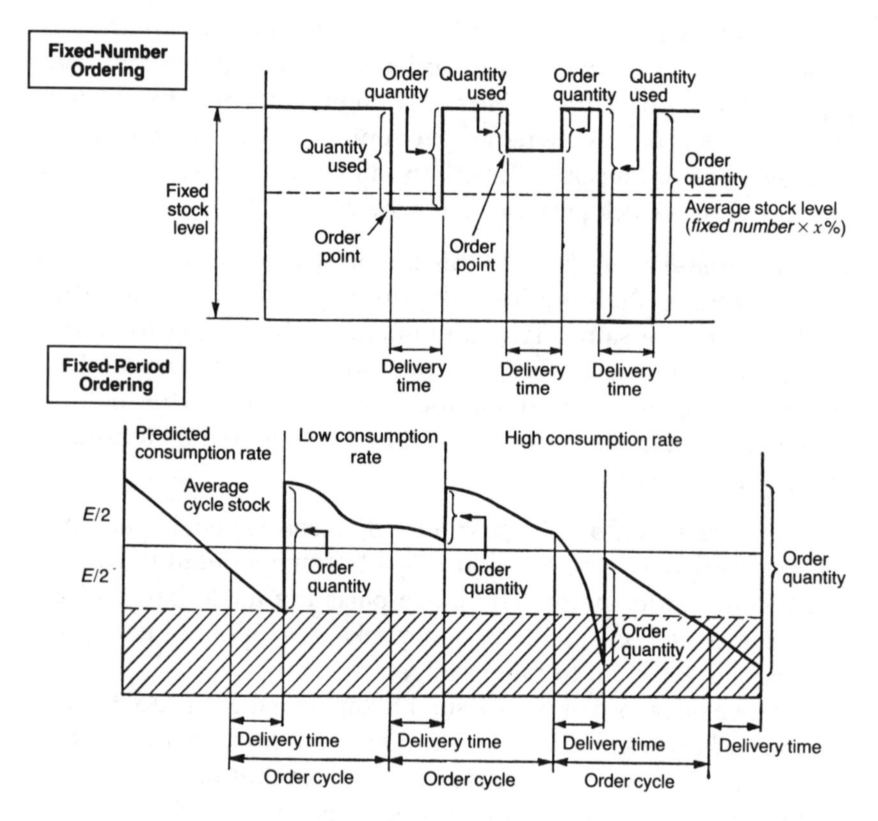

Figure 5-4. Permanent Stock Ordering Methods

Batch issue method. In this method, the person requesting the parts receives a batch to store at the shop floor where they will be used. Another batch is requested when the parts are used up. Batches contain a standard number of parts. This is a suitable method for reducing the administrative costs of handling cheap, frequently used parts such as bolts and nuts.

Fixed-quantity ordering method. In this method, the maximum inventory quantity is set to as small a value as possible, and an order is issued each time a part is used. Thus a fixed number of parts is always kept in stock. This method is suitable for expensive, infrequently required parts such as shafts or other high-priced metal parts.

Fixed-interval ordering method. In this method, the ordering interval is set to a fixed period such as once a year or once every six months. The amount to be ordered varies as necessary. This method is suitable for parts whose frequency of use remains fairly constant, such as tools. It is convenient to coordinate the ordering dates with the financial schedule.

Methods Based on Special Contracts with Suppliers

These methods are based on special contracts with the suppliers to simplify ordering, delivery, and storage, and to make administrative procedures more efficient.

Partial-delivery method based on unit-price contract. Under this method, the unit price of the parts is determined from the average quantity used over the order period. A watch is kept on the stock level, and the parts are delivered in small batches. This is an effective method for permanently stocked parts with a fixed rate of consumption.

Deposit system. In this system, the warehouse is in effect loaned to the supplier, who retains possession of the materials deposited in it. Only the materials used are paid for, which is an effective method for parts with a constant rate of consumption and specifications that rarely change. Because it is so convenient, however, the inventory level tends to increase and storage can become a problem. It is also difficult to switch to another vendor.

Selecting Permanent Stock

Consider each item and size of spare part to see whether it is possible to estimate when and in what quantity it should be used. In principle, if estimation is possible, that part should not be treated as permanent stock. If the part is required often, however, it may be treated as permanent stock to make ordering more efficient.

If the use of a spare part cannot be estimated, it must be treated as permanent stock. While this may result in high inventory costs, the decision must take into account possible production losses due to stoppages. It is usually less costly to keep a permanent stock of the parts whose unavailability will cause serious production stoppages.

Spare parts with the following characteristics should be designated as permanent stock whether or not they can be estimated:

- Parts that must be available in the event of breakdown unless backup machinery or equipment is available.
- Parts that must be purchased three or four times a year. (Those that must be purchased two to three times a year can be planned for and should not be treated as permanent stock).
- Parts that are likely to fail between maintenance periods. These should be treated as permanent stock in readiness for breakdowns.
- Replacement parts salvaged for emergency repairs, such as compressor cylinder valves.
- Parts with delivery times longer than the planned service intervals.

Fixed-quantity ordering is the most common method for maintenance spare parts. When selecting permanent stock items, consider what those who will be affected by the decision have to say and decide the ordering method at the same time.

Establishing Order Points and Quantities

For daily control of permanent stock, two standards are needed: the *order point* (when to order) and the *order quantity* (how much to order).

Using the same method to control all the permanent stock often leads to overstocking. Therefore, order spare parts in the following three ways depending on how the parts are consumed:

Fixed quantity. A fixed quantity is ordered when the order point is reached. The order quantity is based on the economic lot (Figure 5-5).

Fixed number. When parts are used, the number used is ordered immediately, thus keeping a fixed number always in stock. Since this method results in frequent ordering, it is only suitable for expensive, infrequently used parts.

Salvaged parts. These parts are repaired, stored, and reused. If the item is expensive, try to increase the turnaround efficiency.

Issues to be discussed regarding salvaged parts are as follows:

- How many parts, both new and salvaged, should be readily available?
- At what inventory level (new parts + salvaged parts) should an order be issued?
- What quantity should be ordered?

These are calculated by either the fixed-quantity or fixed-number method.

MAINTENANCE BUDGET CONTROL AND MAINTENANCE COST REDUCTION

Equipment budgets are generally classified and controlled according to the purpose of the expenditure or the type of work, either as capital expenditure or running costs or as a period expense. Since capital expenditures (generally categorized in a

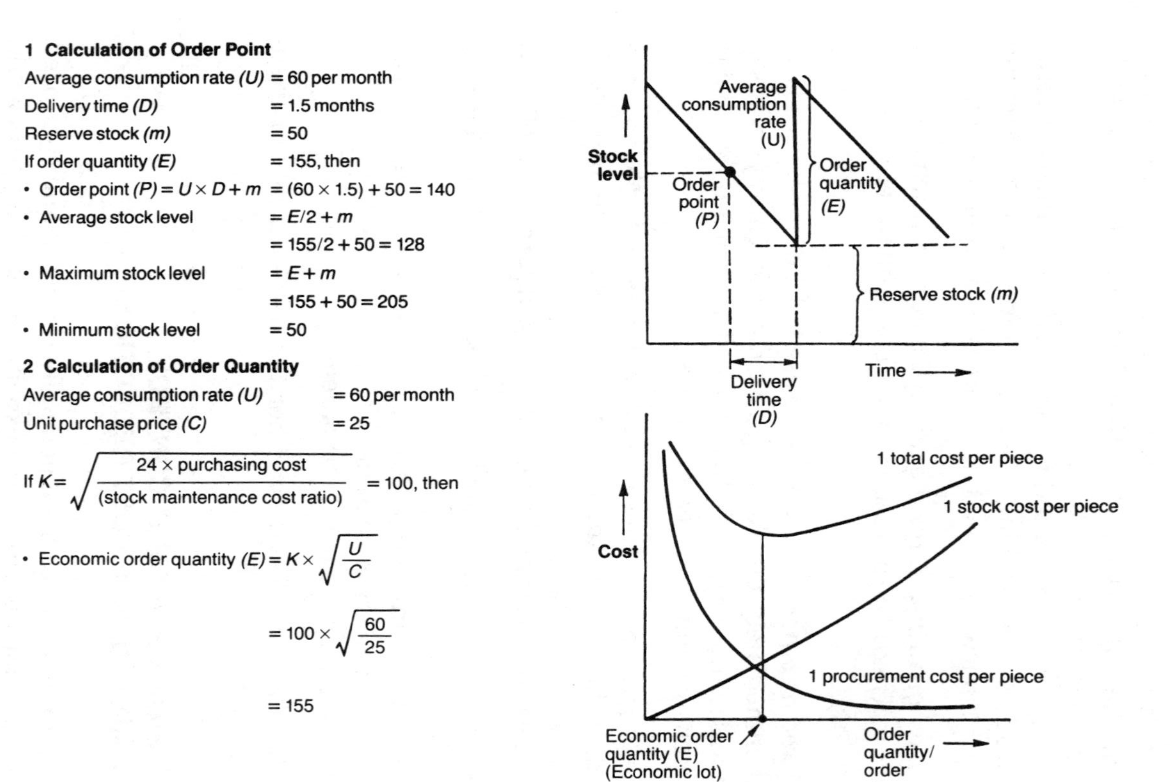

Figure 5-5. Sample Calculations for Fixed-Quantity Ordering Method

fixed-asset account) and running costs (generally expensed) are treated completely differently for accounting purposes, they must be clearly distinguished.

What Are Maintenance Costs?

Capital expenditures, such as the cost of new, expanded, replaced, or revamped equipment, are purchases that are treated as fixed assets. Depending on the company, they are part of a capital-improvement budget, construction budget, expansion budget, or new-product budget.

Running costs, on the other hand, are expenditures that are treated as the cost of maintaining and restoring equipment to its intended use. These are maintenance costs and are generally referred to as overhaul costs, repair costs, restoring costs, and so on.

Maintenance costs are usually treated as running costs from an accounting standpoint and are identified by their grouping into categories such as maintenance material costs, maintenance labor costs, and disbursed maintenance costs (subcontracting costs). To help control the maintenance budget, however, they should be classified in one of the following ways to provide effective control data:

Classification by Purpose

The following costs are classified by their purpose:

Routine maintenance costs include labor and material costs for routine maintenance activities designed to prevent equipment deterioration such as cleaning, lubrication, inspection, and adjustment.

Equipment inspection costs include labor and material costs for inspections to discover abnormalities and determine whether equipment is serviceable or defective.

Repair costs include labor and material costs for repairs to restore equipment to its original condition.

Classification by Maintenance Method

Costs are also classified by the type of maintenance performed:

- preventive maintenance costs (PM)
- breakdown maintenance costs (BM)
- maintainability improvement costs (MI)

Classification by Constituent Elements

Other costs are classified by their elements:

Maintenance material costs cover the costs of materials used for maintenance, such as spare parts, general materials, consumables, lubricants, jigs, tools, and so on. It is important to identify and analyze these items in detail.

In-house labor costs include the labor costs of operators performing autonomous maintenance as well as the labor costs of the maintenance department.

Subcontracting costs are maintenance costs paid to outside contractors.

Other Cost Classification Methods

Other ways of classifying costs include

- *scale of work*, (*e.g.*, large-scale, major maintenance projects, and miscellaneous, small jobs)
- *type of work*, (*e.g.*, mechanical, electrical, piping, instrumentation)

Estimation and Control of Maintenance Budgets

The following cost classifications are helpful in estimating and controlling maintenance budgets (Table 5-12):

	Expenditure	Description	Remarks
1	Spare parts costs	Costs generated when new parts are issued and used	spare parts
2	Parts repair, restoration, and overhaul costs	Costs of processing parts for re-use, *e.g.*, replacing rubber linings, overlaying, machining, and welding	salvaged parts
3	Maintenance labor costs	Inspection, adjustment, repair, parts replacement, and other labor costs	
4	General materials costs	Steel materials, cleaning oil, cotton waste, rubber, paint, seals, and miscellaneous materials costs	
5	Lubricant costs	Lubricating oils, hydraulic fluids, etc.	
6	Maintainability improvement	Costs of accident prevention, lifetime extension, breakdown reduction, and other improvements for maintenance purposes	Maintainability improvement cost
7	Production improvement costs	Costs to improve productivity, such as product quality and yield increases, energy consumption, and so on	Production improvement cost
8	Jig and tool costs	Costs of jigs and tools for maintenance work	
9	Commissioning costs	Costs of repairing design weaknesses and breakdowns in newly-installed equipment. Such problems often occur during the commissioning phase due to problems running in process or lack of familiarity of operators and maintenance staff	These costs are essentially different from normal repair costs, so identify separately as commissioning costs

From Hibi, *Maintenance Economy* (Tokyo: Nikkan Kōgyō Shimbunsha, 1968)

Table 5-12. Classification of Maintenance Costs for Effective Budget Control

Methods of Estimating Maintenance Budgets

Various methods for estimating maintenance budgets are generally used:

Estimate based on actual expenditures. Since maintenance costs, unlike material costs, do not increase or decrease in proportion to operating ratios, they can be estimated on the basis of the previous year's actual expenditures. In this method, changes in operating rates and other conditions are considered and the budget is estimated by slightly adjusting the previous year's figures upward or downward.

Repair-cost rate method. In this method, the equipment cost is multiplied by a maintenance-cost percentage calculated from past expenditures. For example, if the maintenance costs amounted to 6 percent of the equipment acquisition cost, the budget for equipment costing $50,000 would be $\$50,000 \times 0.06 = \$3,000$.

Unit-cost method. In this method, charts are prepared correlating production amounts, operating times, electricity consumed, or other variables, with data on actual repair costs. These charts are then used to calculate the maintenance budget. In the formula $y = ax + b$, y is the budget amount, x is the production amount, operating time, electricity consumed, or other yardstick, a is the maintenance cost per unit of the yardstick, and b is the fixed cost.

Zero-base method. This is a detailed method of accumulating material, labor, and other maintenance costs. The maintenance budget is estimated by reviewing every item of equipment on the annual maintenance plan (the maintenance calendar) and calculating the amount of material and labor needed.

Mixed methods. The four methods described above can be combined as appropriate.

Compiling Maintenance Cost Budgets

Maintenance costs are treated as running costs for accounting purposes and are dealt with separately from equipment budgets, which constitute capital expenditures. For tax purposes, capital expenditures representing fixed-asset purchases must be clearly differentiated from maintenance costs. Technically speaking, however, maintenance work is often a mixture of revamping, replacement, repair, and other work, and is related to equipment budgets for capital expenditure. Therefore, equipment budgets and maintenance budgets should be considered together.

Figure 5-6 shows what happens in practice. Budget allocations are derived from management policies and plans, while budget demands are calculated from totals for major maintenance projects, estimated from historical data. This process is called budget reconciliation. Once budget allocations are reconciled to budget demands, the budget has been compiled.

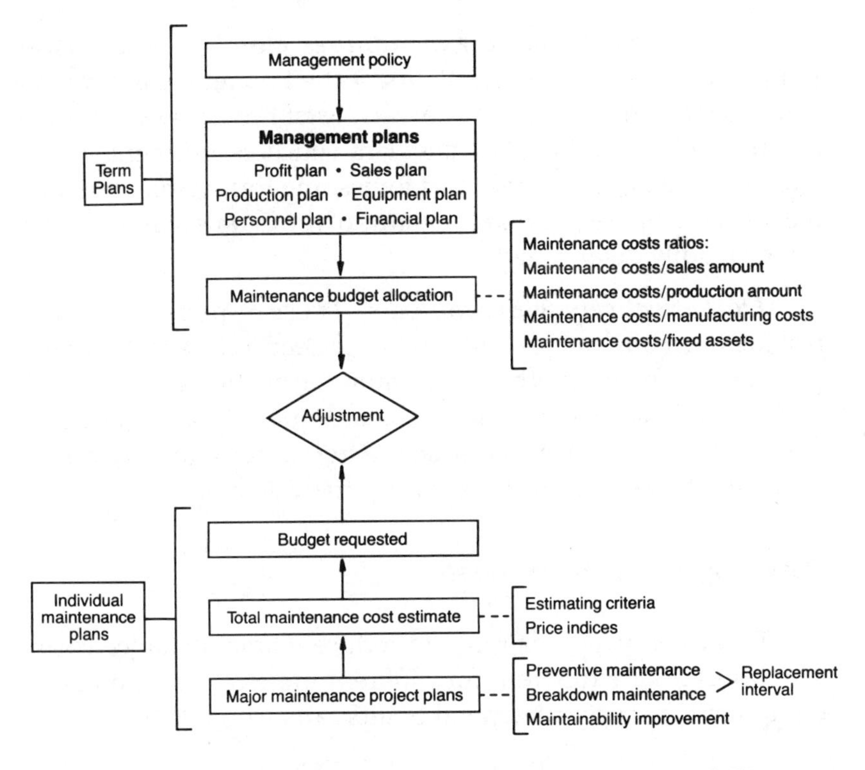

S. Nakajima, *Introduction to Plant Engineering* (Tokyo: Japan Management Association, 1970), 545

Figure 5-6. Maintenance Budget Compilation Flowchart

Maintenance Budget Control

Maintenance budget control means controlling maintenance activities so that the fiscal year (or period) budget targets are achieved. For effective control activity, consider the following points:

Make everyone aware of the need for budget control. Explain the period's maintenance budget and plans to all maintenance department personnel and everyone else concerned. Make certain everyone understands and accepts current industrial trends, the company's industrial position, its customers' requirements, and other background information. Persuade them of the need for budget control and press for their cooperation.

Monitor maintenance expenditures closely. As the fiscal year progresses, review the status of the budget commitments and expenditures at regular intervals. Establish clear systems for issuing, sorting, and totaling payment vouchers indicating actual maintenance expenditures, and for issuing formal maintenance records and reports, so that the status of the maintenance budget can be readily determined.

Deal effectively with problems. Those responsible for controlling the maintenance budget should keep a close watch on its progress. Use the methods for monitoring the status of the budget outlined above to identify and quantify any deviation from the maintenance cost totals. After identifying any deviation, target its source and take appropriate action.

Reducing Maintenance Costs

Every company has room to reduce its maintenance costs. The priorities are different for different industries and types of equipment, but several general points can be noted.

Review Periodic Maintenance Intervals

No two components or parts in a machine deteriorate at the same rate, so periodic maintenance or overhaul intervals should be determined by the parts with the shortest lives. Maintenance is often set at six-month, ten-month, or other scheduled intervals. All intervals should be reviewed at least once a year. It is also important to introduce equipment diagnostic techniques and to switch over, step by step, from time-based to condition-based methods.

Switch from Outside Contracting to In-House Fabrication

The biggest item in the maintenance budget is preventive maintenance work, and disbursed costs usually account for most of it. If too much reliance is placed on outside contractors, valuable in-house maintenance techniques and skills will flow out, making it impossible to develop the necessary PM experience. Subcontracted maintenance should be gradually brought in-house so that all work currently contracted out is eventually performed internally.

Review Spare Parts

A look inside most spare-parts stores will show that inventory levels are higher than necessary. Valves, flanges, V-belts, and gaskets are sometimes stored for two or three years consumption and, in some cases, deteriorate before they are used. Even when order-point control is being exercised, the order point in use may no longer suit present conditions. Reduce the number of permanent stock items and increase the number of planned-purchase items.

Use Idle Equipment Effectively

Longstanding throwaway habits encourage the discarding of old equipment as soon as it has been replaced. Try instead to conserve resources and consider whether an article can be salvaged and reused before discarding it.

Often, plants have many unused pumps, motors, and other equipment. Contact other plants in the company to see if these can be used elsewhere.

Reduce Energy and Resource Use

A tour of any plant usually reveals energy waste. Obvious waste includes leaking oil, compressed air, steam, and water; lights left on unnecessarily; and idling furnaces and motors, for example. Scattered parts, raw materials fallen from the line, and other forms of waste are also easy to find.

Eliminate Equipment Losses

Considerable losses are generated by equipment — for example, energy and yield losses during repair and restoration and product defects due to deterioration in equipment functions. These can be greatly reduced by maximizing equipment efficiency through the introduction of TPM.

Companywide Cost Reduction Activities

Maintenance costs are often hard to reduce in spite of daily efforts by maintenance personnel. More dramatic cost reductions are achieved when operators and other indirect personnel are also involved in companywide activities. These group activities can be organized in the following manner:

Form a project team. Form a project team combining the maintenance, engineering, and production departments with purchasing and accounting staff.

Identify current maintenance costs. Examine the maintenance expenditures for the previous year or period. Establish how much each department spent on each item of equipment, whether the work was done by outside contractors or in-house, and what kind of work was performed. Table 5-13 (*see* pp. 270-271) is an example of a maintenance costs survey form.

Set targets. Set improvement targets after examining the maintenance costs for the whole factory from the totals columns on the costs survey forms. In particular, look beyond outside payments and consider in-house labor costs, especially management-personnel costs.

Prepare progress plans. After setting targets, prepare detailed progress plans showing when individual targets are to be achieved and who will be responsible for them.

Select priority equipment. Prepare pareto analyses for each item of equipment from the maintenance-costs survey charts mentioned above, and designate A-ranked equipment as priority equipment. The number of priority items should be no larger than the number of people on the project team.

Select priority cost items. Select priority cost items from the cost types and uses shown on the survey forms. Determine the order of priority by preparing pareto analyses for each. To maximize effectiveness, select a single item initially and tackle it thoroughly, checking the results before moving on to the next item.

Carry out appropriate improvement measures for each item. Once the targeted equipment and items have been determined, plan and take the appropriate action. Analyze the sources of maintenance costs and prepare improvement plans for tackling them. A good way to produce new ideas is to bring project team members together with others concerned and conduct brainstorming sessions.

Measure results and follow up. As the work is carried out, the project leader should ensure that results are measured and evaluated properly at each step of the process. If only a tenuous connection exists between targets set and actual progress, and if results are not clearly measured and evaluated, little practical effect is achieved.

Follow-up action is also important. For example, decide what should be done about targets that could not be met; find out when such targets can be achieved realistically, and so on.

LUBRICATION CONTROL

Lubrication control ensures that equipment is properly lubricated and that lubrication problems such as leakage and contamination are prevented. Lubrication control increases equipment cost-effectiveness by raising productivity and reducing operating and maintenance costs.

Lubrication control is generally split into control of the lubricant materials themselves and control of the techniques used.

Maintenance Cost Breakdown

Plant: ____________________

Process	Equipment	Year (period) Budget	Year (period) In-house cost and expenditure	Type of Costs			
				In-house labor	Materials/ spare parts	Sub-contracted parts	Stoppage loss
A	a–1						
	a–2						
	a–3						
B	b–1						
Y	y–4						
	y–5						
Z	z–1						
	z–2						
Total							

Ishii, "How to Reduce Maintenance Costs", *Plant Engineer* (February 1978)

Table 5-13. Maintenance Cost Breakdown

Materials control covers the purchase, acceptance, storage, and issue of fresh lubricants and the disposal of spent lubricants; technical control covers the actual use of the lubricants, that is, the type, quantity, method, and timing of application.

Type of Costs		Usage				Remarks
Other	Total	Preventive maintenance costs	Breakdown maintenance costs	Improvement costs	Total	

Types and Uses of Lubricants

The lubricants used in most ordinary mechanical equipment can be grouped in the following three categories, depending on their use: lubricating oils, greases, and solid lubricants.

Lubricating Oils

General lubricating oils. There are many lubricating oils — from mineral-oil-based, animal-based, and vegetable-based to synthetic and compound oils. The main types are mineral-oil-based and compound oils.

Cutting oils. These reduce friction between cutting tools and the materials being machined, prevent the tool from depositing on the work, and reduce tool wear. They increase the machinability of the workpiece, the accuracy of the finished surface, and the lifetime of the tools.

Greases

Greases are lubricating oils mixed with soap or inorganic thickening agents, making them semisolid or semiliquid at normal temperatures. Greases do not drip, splatter, or become contaminated easily. While they adhere well to surfaces to be lubricated and have excellent load-bearing capacity, their disadvantages include considerable mixing resistance and consequent easy heating, poor thermal dissipation and cooling, and large viscosity changes produced by changes in temperature and shearing force.

Solid Lubricants

Various types of solid lubricant have come into use recently. Most of these are used in conjunction with oils and greases, to which they may be added under certain limited conditions. Molybdenum disulfide is a typical example.

Lubricating Methods

Lubricating methods may be classified broadly into disposable, or once-through, methods and self-contained methods. The characteristics and uses of these methods are shown in Figures 5-7 and 5-8.

Lubrication Method	Characteristics	Practical Examples
Hand-lubrication method 	Oil supplied by hand-held oil can	Spinning machines, printing machines, machine tools, and other equipment with low rotation speeds
Sight-feed method 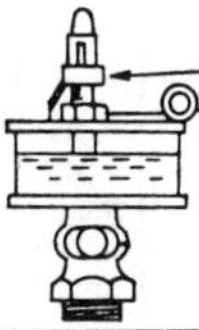 Oil feed-rate adjusting screw Cap	Oil is held in a glass reservoir and supplied at an adjustable rate	Engines, pumps, compressor crosshead guides, machine tools, etc.
Wick-feed method 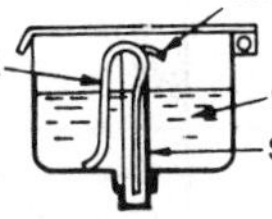 Holder (wire) Wick Oil Standpipe	This device utilizes the syphon action of a wick, drawing up oil from one end and dripping it from the other	Locomotive engines, small electric motors, etc.
Mechanical force-feed method 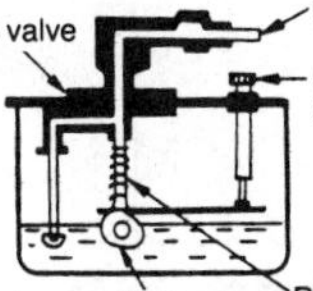 Oil supply pipe Non-return valve Adjusting screw Cam Plunger	Oil dripping down from a sight gage is sent through the oil supply pipe by a plunger pump	Used for steam cylinders, internal combustion engines, and horizontal compressors
Pad method 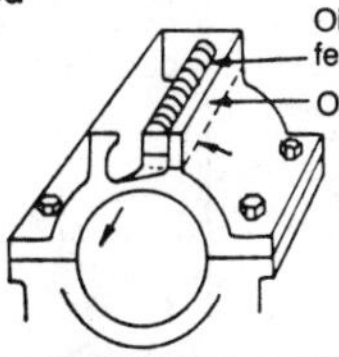 Oil-saturated felt pad Oil basin	Oil seeps evenly through a felt pad from an oil basin above the shaft	Used for open half-bearings
Oil-mist method 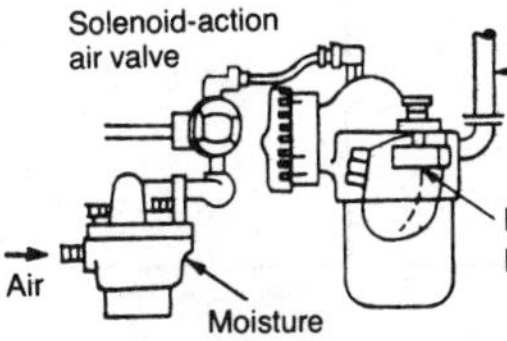 Solenoid-action air valve Oil mist Baffle plate Air Moisture separator	A small amount of oil is mixed with air under pressure forming a mist that lightly coats the contact surfaces	High-speed bearings, plain or anti-friction, enclosed gears, chains, slides, etc.

Figure 5-7. Once-Through Lubrication Methods

Lubrication Method	Characteristics	Practical Examples
Bottom pad method  Packing (woolen yarn)	The packing soaked in lubricant absorbs through capillary action and lubricates the rubbing surfaces by contact	Wheel bearings, etc.
Oil-chain method 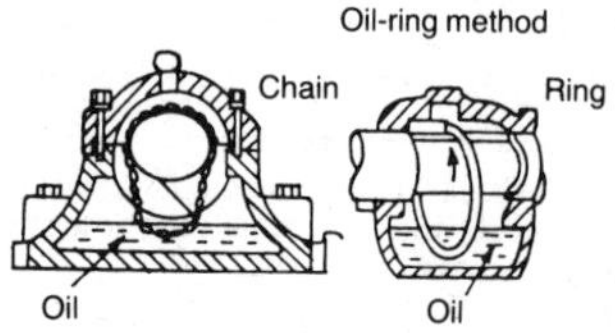 Oil-ring method Chain Ring Oil Oil	Oil is lifted up and supplied to the contact surfaces by a rotating ring or chain	Crankcases of low-speed, intermediate or high-load compressors, pillow blocks, general machinery bearings, etc.
Worm-gear oil bath method 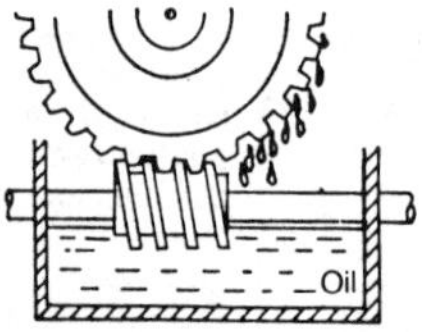 Oil	Oil is lifted up to the rubbing surfaces by the rotating action	Low and intermediate-speed sealed gears, speed reducer bearings, etc.
Circulating method 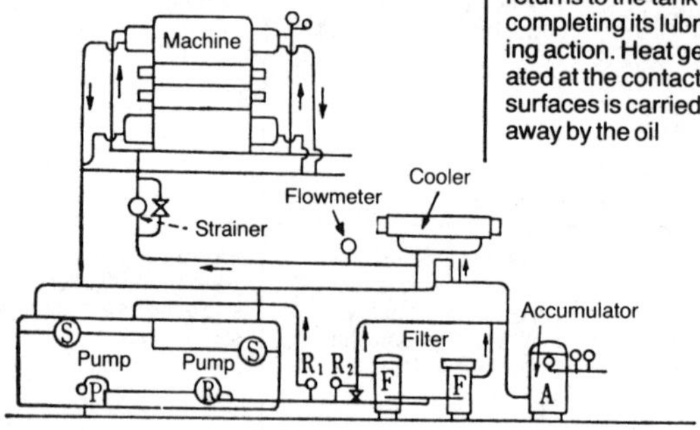 Machine Cooler Flowmeter Strainer Accumulator Filter Pump Pump S S R1 R2 P R F F A	The oil is sent from the oil tank to the contact surfaces by pumps and returns to the tank after completing its lubricating action. Heat generated at the contact surfaces is carried away by the oil	Steam turbines, diesel engines, bearings of rolling mills, screens, speed reducers, and other large machines

Figure 5-8. Self-Contained Lubrication Methods

Once-through Methods

Hand lubrication. In this method oil must be added at short intervals because too much oil is present just after application and too little after time has passed. The method should be restricted to high-viscosity oils and used where there is little movement.

Sight-feed oiler. In this method oil is drop-fed by gravity. The amount of oil is proportional to the aperture of the feed valve and to the square root of the valve's distance below the surface of the oil in the supply tank. Thus the rate at which the oil is supplied varies with time. For example, if the tank is refilled when the oil level has dropped by 40 percent, the drip rate will be two-thirds its original value. The rate is also easily affected by vibration and changes in the atmospheric temperature, however, and valves are easily clogged by dust or other foreign matter. For this reason, the equipment used in this method requires careful maintenance.

Syphon-type wicker oiler. Since this method makes use of the siphon action of a wick, the supply rate varies with changes in the oil level even more than in the sight-feed oiler. The same care should be taken in maintaining the equipment used in this method. This method is unsuitable for high-viscosity oils.

Mechanical force-feed lubricator. In this method, the supply of oil is controlled by varying the stroke of a plunger with an adjusting screw. The lubricator can be checked during operation by means of a sight tube. If no oil drops can be seen, the cause may be wear of the plunger cam, blockage in the oil supply pipe, or a low supply of oil. When this occurs, the unit must be dismantled and inspected immediately.

Pad oiler. In this method, even lubrication is achieved through a felt pad in the oil reservoir that also acts as a filter. However, when there is only a small amount of oil present, the felt tends to harden at the contact points and wear away, leading to insufficient lubrication. The equipment should therefore be inspected frequently.

Oil mist. This method injects oil drop by drop in a stream of pressurized air, creating a mist that applies just enough oil to wet the operating surfaces. It has various advantages and has recently been widely adopted. The equipment has relatively few parts and is easy to maintain. If chosen to match the equipment, it is an efficient method of lubrication. Only the air carrying the oil must be controlled.

Self-contained Methods

Bottom-feed wick oiler. In addition to protecting the equipment from dust, the packing helps to filter the oil. It must therefore be cleaned periodically.

Ring oiler. In this method, the oil is circulated and reused over relatively long periods and should therefore be of intermediate quality or higher. Take care when using high pour-point oils at low temperatures, since their flow resistance will be high and this will check the rotation of the ring or chain and increase the likelihood of insufficient lubrication.

Worm-gear bath oiler. Here, the oil is recycled over a long period and inevitably becomes contaminated by metal particles from the contact surfaces and by dust and other foreign matter from outside. It should therefore be analyzed periodically and replaced as necessary. Also, check the oil level as part of routine inspection procedures.

Pressure-circulating system. When this method is used to supply oil to many lubrication points on large items of machinery, insufficient lubrication will have very serious consequences. For this reason, and because the same oil is circulated over long periods of time, use an oil with excellent antioxidation, antifoaming, anticorrosion, and moisture separation properties. Analyze the oil regularly and recondition it, replace it, or adjust its viscosity at suitable intervals. The circulating equipment itself must also be carefully maintained. In addition to checking oil gages, cleaning and inspecting filters and strainers, and dismantling and inspecting coolers and heaters to prevent any loss of efficiency, routinely inspect all oil supply rates, levels, and pressures.

In self-contained systems of lubrication, the lubricant is in constant contact with metal surfaces over long periods and is affected by air, moisture, heat, and the like, each of which contributes to deterioration of the lubricant. It turns brown, its viscosity increases, and eventually it gives off a pungent odor and loses its lubricating properties.

Controlling Deterioration and Contamination of Lubricants

Lubricants deteriorate or become contaminated for a variety of reasons that must be well understood if lubrication is to be effectively controlled.

Oxidation

A lubricant deteriorates because its unstable constituents absorb oxygen from the air and form oxides. The following factors accelerate this process:

Heat. The rate of oxidation generally doubles when the temperature rises by 10° C.

Contact with metals. All lubricants exercise their lubricating action by contact with the surfaces of metals. The lubricants become contaminated by metal particles produced by friction. Since these particles are extremely small, they present a large surface area to the lubricant and promote oxidation through catalytic action.

Contact with moisture. If moisture is present in a lubricant, it forms an emulsion when agitated, initiating corrosion of the metal surfaces and promoting oxidation of the lubricant.

Contamination

Lubricants become contaminated both by metal particles from the contact surfaces and by foreign matter from outside. In practice, the effect of this contamination is greater than the effect of deterioration of the oil through oxidation. Contamination produces seizure, scuffing, abrasion, cavitation, fatigue of contact surfaces, noise, vibration, deterioration of the lubricant, and so on. The contamination may consist of a fine residue of sand, carbon, metal, fiber, or other materials. This is an extremely troublesome problem, since 70 to 80 percent of these particles are less than 10μ in diameter, while particles visible to the naked eye are at least 100μ in diameter.

Although the rates are different for different types of lubricants, they all gradually deteriorate and become contaminated. Methods of combating this are summarized in Table 5-14. The points listed in the table should be observed carefully and efforts made to preserve the lubricant's properties over long periods.

1. Use oils with viscosity and other properties suitable for the operating conditions. Under severe operating conditions, use high-grade oils with appropriate additives.
2. Avoid unnecessary agitation caused by oversupply of oil and, as much as possible, keep oils away from contact with air.
3. Analyze oils regularly during use and periodically recondition them or adjust their viscosity to avoid hastening deterioration.
4. Do not raise the temperature of the lubrication system unnecessarily.
5. Clean the lubrication system regularly and periodically remove contamination from within the system.
6. To prevent the introduction of moisture, steam and dust, use appropriate sealing devices and covers and ventilate the lubrication system.
7. Use lubrication devices suited to the equipment and maintain them scrupulously.

Table 5-14. Actions to Prevent Deterioration and Contamination of Lubricants

Key Points for Daily Inspection

No matter how high the quality of the lubricant and lubricating devices used, they should be maintained and used under optimum conditions. It is vital to detect abnormalities quickly and take the right corrective action. Key points for routine inspection of lubricants are listed below:

Control lubricant level. Checking levels is one of the most important points in maintaining proper lubrication. Set appropriate levels for each item of equipment and take steps to ensure that the levels established can be properly maintained. Although conditions of application vary, general standards for the different lubrication methods are shown in Table 5-15.

Check lubricant temperature. Since increases in temperature reduce the viscosity of a lubricant and accelerate its deterioration, rigorous routine checks are needed to ensure that

Oil Level / Lubrication Method	Maximum Oil level	Minimum Oil Level
Drip method	Full height of oil reservoir	⅓ of height of oil reservoir
Wick method	" "	" "
Pad Method	Full height of oil basin	Minimum height which will completely immerse pad
Ring-oiler	⅕ of ring diameter above bottom of ring	⅐ of ring diameter above bottom of ring
Chain-oiler	Maximum height at which no oil will spill from oil basin during operation	Height which fully immerses lower part of chain in oil
Collar-oiler	Bottom of axle	½ height of collar from bottom of axle
Oil-mist method	Strictly observe level specified on oil supply receptacle	
Mechanical force-feed method	Full height of oil reservoir or level gage	⅓ of height of oil reservoir or level gage
Oil-bath method • Compressor crankcases	Level that immerses a suspended weight to a depth of 3–5cm without any oil touching the stuffing box	
• Roller bearings: bm n<8,000	No more than ⅒ of ring diameter above bottom of ring	
• Roller bearings: bm n>8,000	½–⅘ of roller diameter above bottom of rollers	
• Chains: sprocket bath	Height that completely immerses one sprocket tooth	
• Chains: chain bath	Height that immerses ring plate of lowest part of chain	
Pressure-circulating method	9/10 of tank height when stopped	½ tank height when in operation

Table 5-15. Oil Level Standards for Different Lubrication Methods

temperatures do not rise above the specified values or remain there for long periods.

Control lubrication rate. Applying the correct amount of lubricant is the most important aspect of lubrication control. Insufficient lubricant naturally means insufficient lubrication, but excess lubricant is also undesirable because spillage can also be a problem. Make sure the specified amount of lubricant is being supplied to the equipment as part of routine inspection procedures. This must be done very conscientiously, since if lubricant is not being supplied at the proper rate, mechanical wear or breakdown will occur in a very short time.

Lubrication control means ensuring that the correct amount of a suitable lubricant is supplied to the required parts of the equipment in the exact quantity needed. It is important to instill the habit of first checking the state of lubrication whenever any problems are encountered with equipment.

PREDICTIVE MAINTENANCE AND MACHINE DIAGNOSTIC TECHNIQUES

The maintenance methodologies known as *predictive maintenance* and *condition-based maintenance* are attracting attention as highly reliable replacements for conventional periodic maintenance and overhaul. Gradually, more and more Japanese factories are beginning to use these new methods.

The Need for Predictive Maintenance

The methods constitute a new type of preventive maintenance that uses modern measurement and signal-processing techniques to accurately diagnose the condition of equipment during operation and determine when maintenance is required. To remain competitive, companies must switch from periodic maintenance to predictive maintenance for equipment that is expensive to repair or that causes serious losses if it breaks down.

Machine Diagnostic Techniques

The intervals for conventional periodic maintenance and overhaul are usually decided by determining the maximum operating times from breakdown statistics and from visual inspections. As modern equipment has become more advanced and complex, however, its performance can no longer be judged through conventional intuition and experience based on the senses. Moreover, as equipment systems grow in size and complexity, the numbers of widely varying operating conditions also increase. Maximum operating times calculated from breakdown statistics and probability computations are increasingly subject to large experimental errors.

If possible, overhaul and maintenance intervals should be derived scientifically, based on an accurate grasp of machine conditions. This is where the need for machine diagnostic technology comes in.

Machine diagnostic technology measures the stress on equipment and its malfunctions, deterioration, strength, performance, and other properties without dismantling. It helps diagnose and predict equipment reliability and capacity by distinguishing and evaluating the cause, location, and degree of danger of any malfunction, and indicating a method of repair.

Equipment diagnosis naturally centers on identifying current conditions, but it is not simply a question of determining present symptoms. It is a technology for monitoring *continuous change*.

Application and Aims of Predictive Maintenance

The types of breakdown to which predictive maintenance is applicable are limited to those in which changes in previously set parameters can be detected and used to forecast breakdowns. It is not suitable when there is no means of detecting malfunctions in advance. It is also unsuitable when the costs of monitoring will be higher than the savings in repair costs or production losses.

Predictive maintenance aims

- to reduce breakdowns and accidents caused by equipment
- to increase operating times and production
- to reduce maintenance times and costs
- to increase the quality of products and services

Condition Monitoring Techniques

Condition monitoring is the application of machine diagnostic technology. The following seven techniques are used:

Thermal methods. These include the use of thermal paint to render motor overheating visible and thermography to monitor the temperature of furnaces and the condition of power lines.

Lubricant monitoring. Methods range in sophistication from monitoring lubricant color, oxidation, and metal particle content to spectrochemical analysis.

Leak detection. Leaks from pressure vessels are detected using ultrasonics or halogen gases.

Crack detection. Cracks are detected using magnetic flux, electrical resistance, eddy currents, ultrasonic waves, or radiation.

Vibration monitoring. Shock pulse and other methods are used, mainly on machinery with moving parts.

Noise monitoring. Various types of devices monitor the condition of equipment through the noise it generates.

Corrosion monitoring. Acoustic emission and other methods are used to monitor the condition of metals.

Of the above seven methods, thermal monitoring, lubricant monitoring, and vibration monitoring constitute the most frequently used condition-monitoring techniques related to predictive maintenance.

Condition Monitoring and Machine Diagnosis

The condition-monitoring techniques considered suitable for monitoring the general condition of a machine using a number of measuring points are vibration monitoring, thermal monitoring, and lubricant monitoring. These correspond to monitoring a person's pulse, temperature, and blood pressure.

Machine diagnosis starts with quantitative measurement of the machine's condition during operation. Condition monitoring is extremely important, and the key to the rapid detection of malfunctions is selecting the correct technique. In rotating and reciprocating machinery, for example, vibration and noise monitoring are the most effective condition-monitoring techniques. Even under normal conditions, rotating machinery generates vibration and noise because of slight mechanical irregularities or the properties of fluids being processed. This vibration and noise will often increase, however, as a minor internal malfunction develops into a major problem.

Therefore, the vibration and noise levels of the equipment are measured under normal conditions, and the changes in the levels are measured periodically. This reveals the onset of malfunctions and permits the prediction and prevention of deterioration and breakdowns. Moreover, since particular malfunctions generate unique patterns of vibration and noise, an analysis of these will help identify the location and cause of the malfunction.

Case Study 5-1 — Predicting Life Span through Vibration Monitoring

At Fuji Photo Film's Ashigara plant, the life span of blowers was predicted through the use of vibration meters. Vibration meters are useful for diagnosing rotating parts and are generally available.

One thousand blowers were targeted for preventive maintenance. Since they operated 24 hours a day, a breakdown seriously affected the associated equipment. Repairs took as long as five or six hours. When the preventive maintenance system was introduced, the blowers were overhauled once every two years. No malfunctions were detected in most of the blowers during these overhauls, however; on the contrary, many malfunctions were actually caused by the overhauls. After considering whether to extend the overhaul intervals, the project team decided to measure the deterioration of the blowers and use vibration meters to help judge when overhauls were needed. This was accomplished after two years of experiments and trials.

Figure 5-9 shows the measurement points on the blowers. At first, the acceleration (*g*) and the vibration amplitude (μ) of the specified points were measured at six points — A through F — giving a total of twelve measurements. As the frequency of measurement increased, adding even one measuring point resulted in a large number of measurements. For this reason, and to reduce the number of measuring manhours required, the measurement points were later reduced to two (A and B), and to A only for less important blowers.

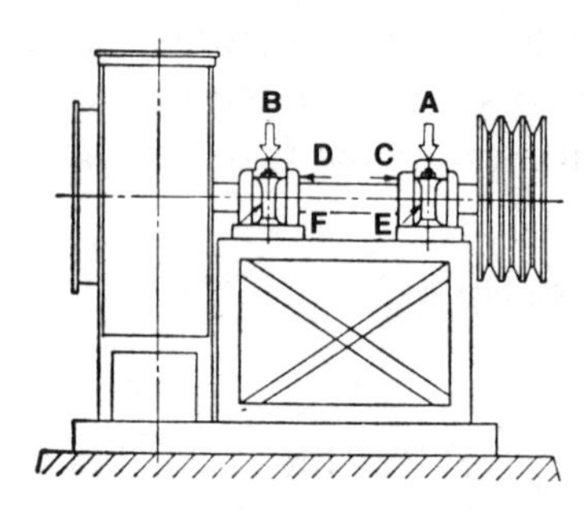

Kagiwada, "Predicting Equipment Lifetimes through Vibration Measurements", *Plant Engineering* (July, 1974)

Figure 5-9. Measurement Points

Figure 5-10 shows the trends in the vibration measurements.

- Blower A is normal
- Blower B: During routine inspections, Blower B was found to be vibrating more than normal. On measuring

the vibration, only the acceleration was higher than before, with a value of 0.6 *g*. Ten days later it reached 1.0 *g*. Overhauling the blower disclosed that the bearing and the shaft on the pulley side were rubbing. The shaft was worn by approximately 5⁄100 of its diameter.

- Blower C: Routine inspection revealed that Blower C was vibrating abnormally in the same manner as Blower B.

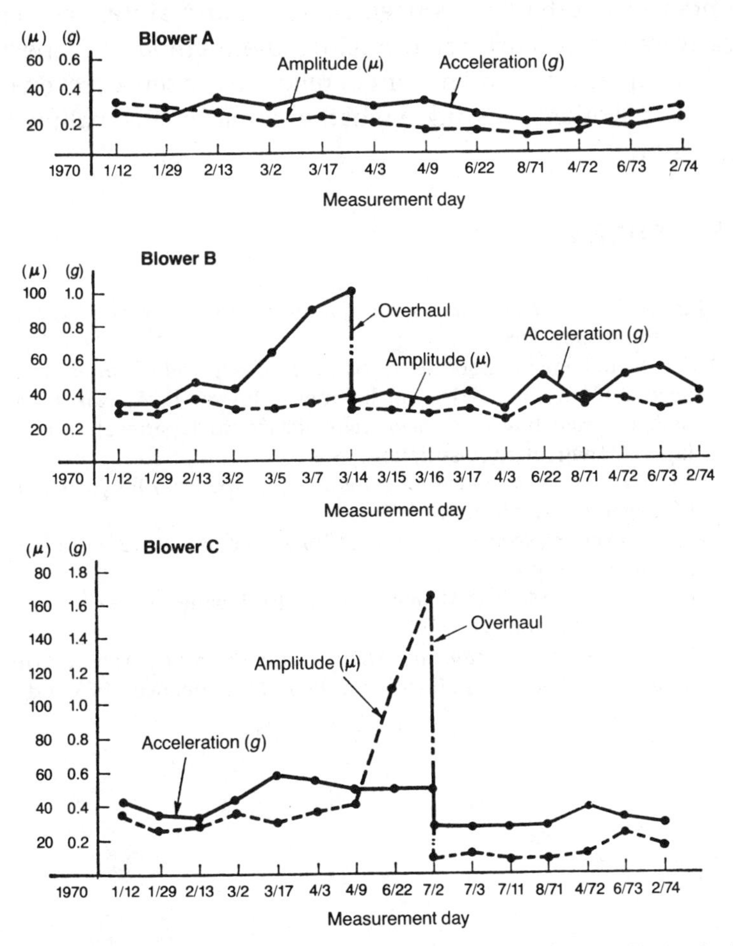

M. Kagiwada, "Predicting Equipment Life through Vibration Measurements", *Plant Engineering* (July, 1974)

Figure 5-10. Results of Vibration Measurements

> Measurements indicated that the vibration amplitude was higher than normal, at 85μ, while the acceleration had not changed. Overhauling the blower revealed that the bearings were normal but the attachment bolts were loose.

The use of vibration meters to predict blower life as described above allowed a switch from conventional time-based maintenance to condition-based maintenance. As a result, the number of overhauls has been halved and maintenance labor greatly reduced. Furthermore, since the method of inspection has been quantified and no longer relies on the senses, even inexperienced workers can judge whether a machine is malfunctioning or not.

REFERENCES

Hibi, S. *Maintenance Economy.* (in Japanese) Tokyo: Nikkan Kōgyō Shimbunsha, 1968.

Japan Management Association, ed., *Basic Equipment (Facility) Maintenance (Course and Text).* (in Japanese) Tokyo: JMA, Nakajima, Seiichi. *Promoting Equipment Maintenance* (in Japanese). Tokyo: Japan Management Association, 1969.

_____ . *Introduction to Plant Engineering.* (in Japanese) Tokyo: Japan Management Association, 1970.

_____ . *Terotechnology.* (in Japanese) Tokyo: Japan Institute of Plant Maintenance, 1975.

Ota, F. *Practical Machine Maintenance I-II.* (in Japanese) Tokyo: Technology Evaluation, Ltd., 1975.

Takahashi, Giichi. *Production Maintenance Promotion Manual.* (in Japanese) Tokyo: Japan Institute of Plant Maintenance, 1975. n.d.

6

Maintenance Prevention

WHAT IS MAINTENANCE PREVENTION?

Equipment management can be roughly divided into project engineering and maintenance engineering. *Maintenance prevention* (MP) is a significant aspect of project engineering that serves as the interface between project and maintenance engineering.

The goal of maintenance prevention activities is to reduce maintenance costs and deterioration losses in new equipment by considering past maintenance data and the latest technology when designing for higher reliability, maintainability, operability, safety, and other requirements. In other words, it means designing and installing equipment that will be easy to maintain and operate.

In the model illustrated in Figure 6-1, equipment engineering is systematized using the following four subdivisions:

1. Equipment investment planning (techniques for evaluating the economics of equipment investment)
2. Early equipment management (MP design technology)
3. Operation and maintenance (technology for maintaining and improving existing equipment)
4. Rationalization measures (technology for equipment development and modification)

Note that MP activities are positioned here as *early equipment management* (from design to commissioning; *see* Figure 6-1.)

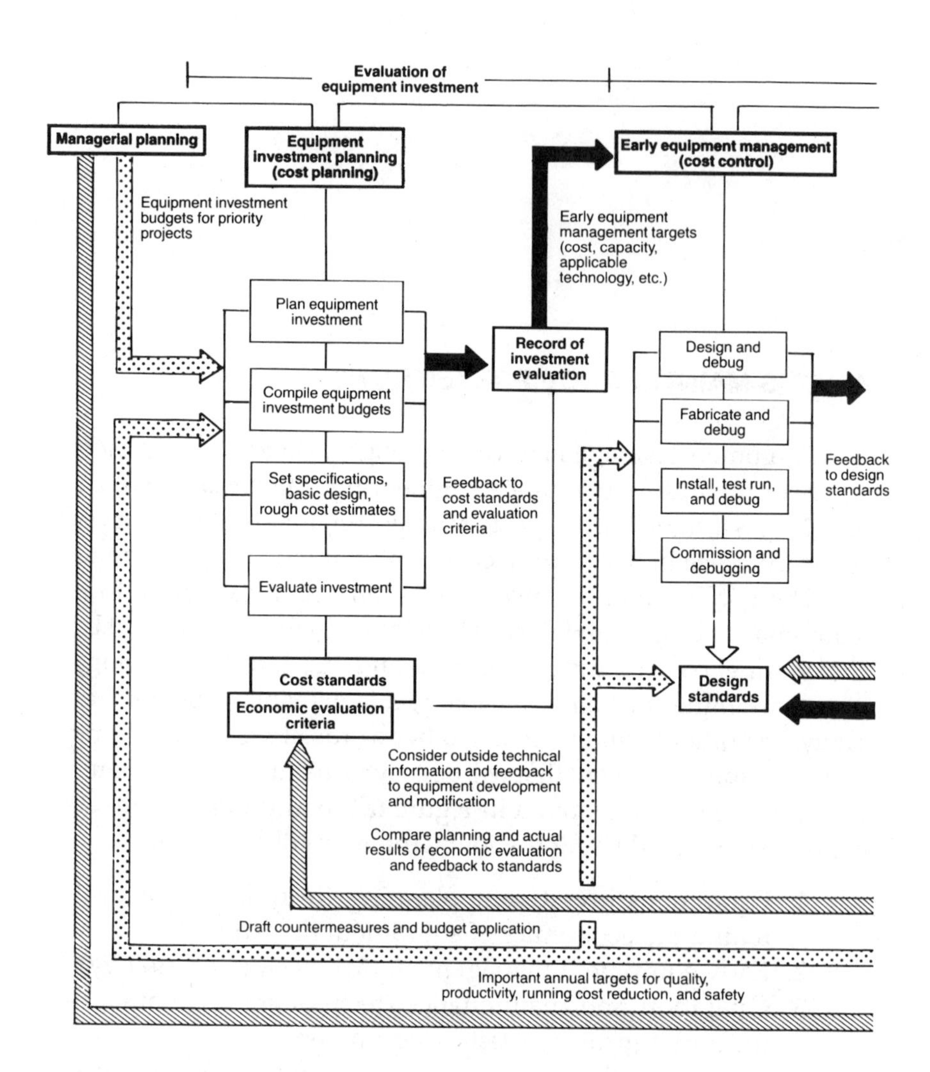

Figure 6-1. Equipment Technology Outline

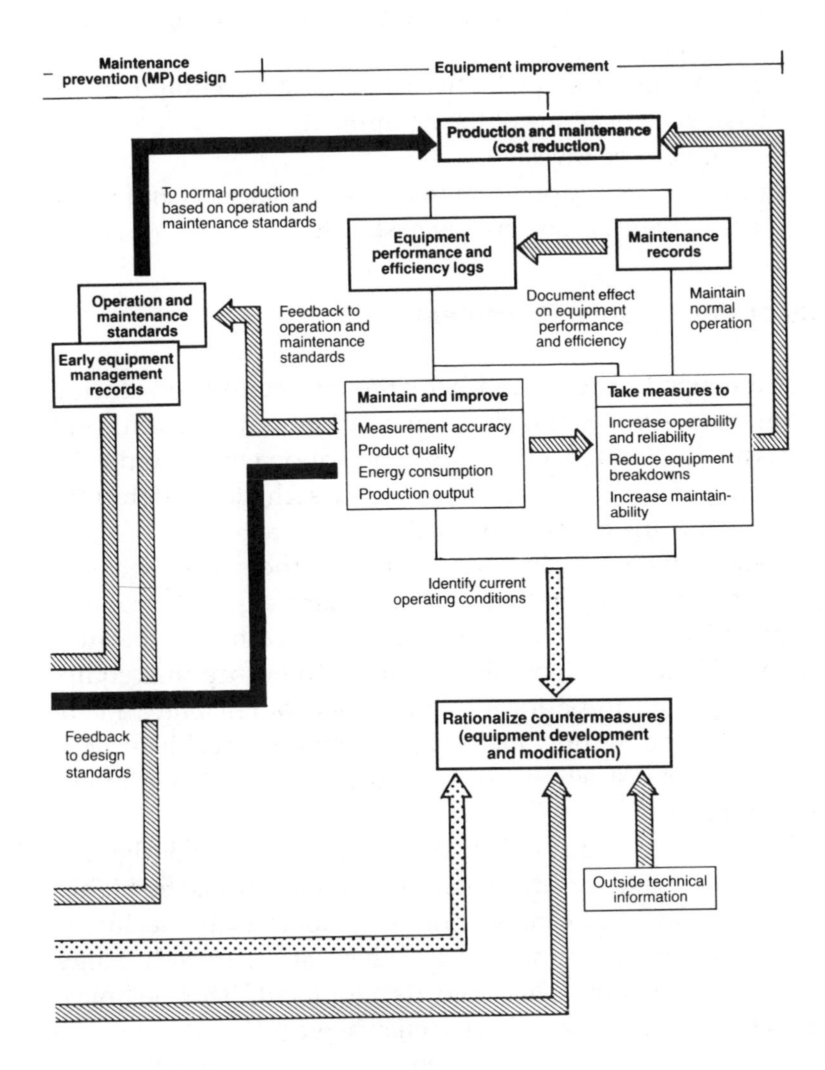
Maintenance prevention (MP) design
Equipment improvement
Production and maintenance (cost reduction)
To normal production based on operation and maintenance standards
Equipment performance and efficiency logs
Maintenance records
Operation and maintenance standards
Early equipment management records
Feedback to operation and maintenance standards
Document effect on equipment performance and efficiency
Maintain normal operation
Maintain and improve
Measurement accuracy
Product quality
Energy consumption
Production output
Take measures to
Increase operability and reliability
Reduce equipment breakdowns
Increase maintain-ability
Identify current operating conditions
Feedback to design standards
Rationalize countermeasures (equipment development and modification)
Outside technical information

At the equipment investment planning stage the following restrictions and aims are considered and decided:

- technology to be used (production technology, equipment technology)
- quantitative and qualitative capacity of the equipment
- basic specifications of the equipment
- amount to be invested
- running costs (operation manpower, materials and yield, maintenance costs, energy costs, etc.)

Aims of Maintenance Prevention

Maintenance prevention activities are conducted during equipment design, fabrication, installation and test run, and commissioning (establishing normal operation with commercial production). They include debugging at each stage (detecting and correcting errors and malfunctions) (Figure 6-2).

These activities are intended to reduce the period between design and stable operation and to assure efficient progress through this period with minimum labor and without imbalance in the workload. They are also intended to ensure that equipment is designed to perform at high levels of reliability, maintainability, economy, operability, and safety, and to achieve these aims within the restrictions set out at the equipment investment planning stage.

To achieve the aims of MP, the engineering and design engineers responsible for equipment development must be highly skilled. They must also make full use of all available technical data, combining the application of this data with technology based on in-house research and development. This technical data includes operating and maintenance records, records of improvements to existing equipment, equipment development and modification records, outside technical data, records of past MP activities, and design standards and checklists based on these data. Analyzing and applying this information should lead to new equipment that requires less maintenance and produces higher quality products.

Figure 6-2 shows the MP design methodology used at Tokai Rubber Industries.

Why Is MP Important?

Without maintenance prevention activities, problems emerge when new equipment is installed during test-run and commissioning, even if design, fabrication, and installation appear to have gone smoothly. Normal operation is difficult to establish, and production and maintenance engineers may have to make many changes before achieving full-scale operation.

Even after the equipment is operating normally, minor repairs and inspection, adjustment, lubrication, and cleaning to prevent deterioration and breakdowns are so complicated that everyone involved becomes thoroughly discouraged. Under such circumstances, inspection, lubrication, and cleaning may be neglected; equipment downtime is prolonged for no good reason, even over very minor breakdowns.

These kinds of trouble often occur at the startup stage, and subsequent equipment modifications consist of tying up loose ends left at the design and fabrication stages. If these phenomena were truly inevitable in the process of increasing scale, speed, and automation, dealing with them would be far too problematical. The role of MP is to minimize these problems, however, by designing safeguards and countermeasures into the equipment before its fabrication and installation.

BRIDGING THE INFORMATION GAP

The quality of subsequent productive maintenance is determined largely by whether the latest technology of equipment reliability and maintainability is brought in from outside or developed in-house through the efforts and experience of production, design, and maintenance staff. It is also determined by whether full use is made of the company's accumulated technical expertise and the exhaustiveness of in-house research and investigations.

The quality of a company's MP program depends on the following three factors:

- technical skills and design sense of the engineering and design engineers
- quality and quantity of technical data available
- ease with which this technical data can be used

Increasing Technical Skill

An increase in engineering ability is developed through the engineer's own creativity and effort, but it is unproductive for the individual to rely entirely on theoretical study and pure engineering experience. Engineers from equipment-related departments should gather technical data from actual operating conditions for use in reliability and maintainability design. This data can be collated and developed into engineering and design standards.

Increasing Availability of Technical Data

Far too often, however, the technical improvement data obtained by maintenance engineers from routine PM activities is not used in reliability and maintainability design. The maintenance engineers do not present this data in a form that is acceptable to design engineers. Moreover, design engineers do not standardize general technical and maintenance data, so the maintenance department cannot apply it.

More effective communication between the maintenance and design engineers is the obvious first step. Maintenance engineers should consider how to compile useful MP data to support the design department in planning and designing their equipment. In return, design engineers should be responsible for equipment they have designed, even after it is fabricated and installed. They should follow up with the understanding that tomorrow's technology will be developed from today's mistakes.

Collecting and Using MP Data

Figure 6-3 shows a method for collecting and standardizing MP data. Safety, quality, maintenance, engineering, and other types of data are gathered, analyzed, and codified. The data is arranged chronologically and stored in equipment or line history files. Common technical data that can be used in many types of equipment is accumulated and standardized in the form of equipment design standards and safety standards. The data is available for use throughout the company.

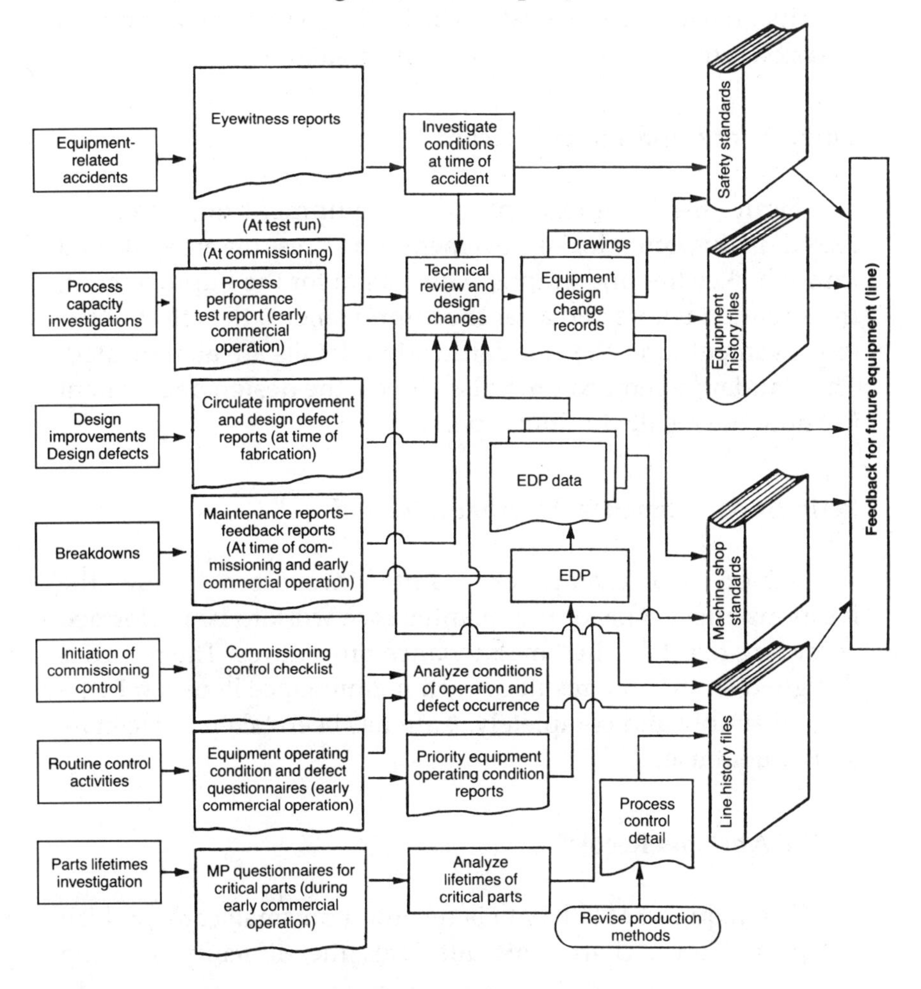

Figure 6-3. Feedback and Standardization of MP Data (Nippondenso)

Classifying Maintenance Data

Maintenance data is a short title, but it covers a large variety and quantity of data, including equipment improvement records, breakdown repair records, periodic maintenance records, precision inspection records, and so on. The engineering department cannot use this data easily in its original form. It must be put in a form that is useful for design, so data can be retrieved when it is needed. For breakdown repair records, for example, consider using hole-sort cards or other record-keeping methods that place priority on use and allow data to be retrieved through key words describing the location or cause of the breakdown.

Equipment Improvement Record

Figure 6-4 is an example of an equipment improvement record. In this example, improvements on a press die are explained through sketches and graphs. The reason for the improvement, the situation before and after the improvement, the effect of the improvement, and the standardization details are all indicated. The standardization column also directs the design department to apply the results to similar parts.

Periodic Maintenance (Overhaul) Record

Table 6-1 is an example of a maintenance record. It shows the locations and details of the maintenance work to be performed according to established maintenance procedures. The record is designed so that the result of each maintenance item can be recorded simply and completely. Care has been taken to facilitate subsequent use.

MTBF Analysis Record

The required MTBF data is recorded on small cards, which are posted on a chart. This alleviates the difficulty of using

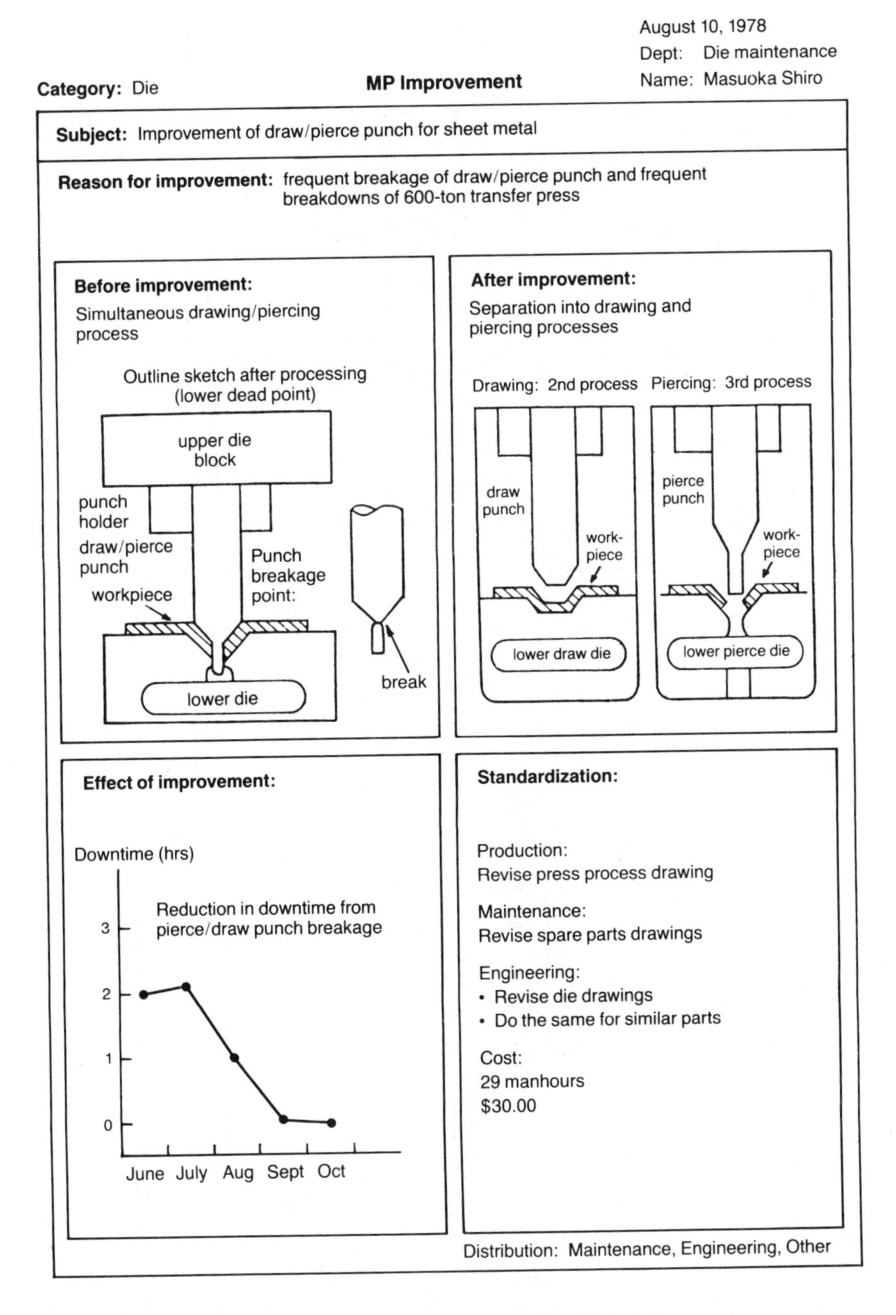

Figure 6-4. Example of MP Improvements (Toyota Steel Works)

Maintenance Record **Dept.: 532** **Line: 14** **Equipment #: M5-026 H No. 4 punch press**

Machine part		Maintenance item	Results	Action and remarks
Section name	Unit name			
	Flywheel	Inspection and checking	○	Replace bearing on flywheel side
	Rotor seal	Inspection and checking	○	
	Safety cover	Inspection and checking	○	

How to complete "Results" column

When maintenance item is "inspection":
- ○: Good (guaranteed until next inspection)
- ×: Defective (includes items which cannot be guaranteed until next inspection)

When maintenance item is anything other than "inspection":
- Execution "completed" or "not completed"
- For checking, record measured values

The aim is to eliminate all defects through the maintenance work indicated on this form. If the work cannot be completed due to lack of time or other reasons, the person responsible for maintenance must indicate the action to be taken by issuing a work order clearly detailing the defects. The work order issue date must be recorded on this form.

Date Issued: April 10,1978
By: Adachi Yasuo (person responsible for maintenance)

Table 6-1. Sample Maintenance Record

maintenance records. Breakdown details can be seen at a glance and problems and improvement priorities are easily identified.

Feedback Record During Equipment Fabrication

Table 6-2 is a sample form for feedback to equipment design, recording the details of trouble encountered and the action to be taken.

MP DESIGN STANDARDIZATION

When new equipment is designed or existing equipment is modified, reliability and maintainability requirements cannot be effectively incorporated if technical data is not properly collated and communicated. Maintenance data filed away before it is put into usable form is worthless; it cannot be considered accumulated technology. Inexperienced engineers cannot increase their technical skills if the technology they require exists only in the heads of the more experienced staff.

The data and technical know-how representing a company's accumulated experience must be standardized and put in the form of guidelines to increase designers' technical skills and prevent mistakes.

The Need for Parts Standardization

One of the most annoying obstacles to efficient maintenance work is overabundance of parts in use. Multiple parts with different designs from different manufacturers are used even when they have identical functions. This not only increases parts inventories, it also results in repair mistakes and long stoppages when parts run out.

Since the use of multiple parts is sometimes the result of designers' enthusiasm, it cannot be condemned out of hand. "Catalog mania" or novelty-hunting should be discouraged, however. Standardize parts wherever possible to halt their proliferation.

Type of record	☐ Feedback to equipment design and action record	(Issue) (records file) (originals file) (Route to responsible design section → planning section)
	☐ Action to correct problems during equipment fabrication	(Route to responsible design section) (records file)

Machine no.*	Machine name*	Drawing no.	Processed material*	User dept.*

No.*	Date rec'd	Dept. supplying data (A)*	Priority (B)*	Problems/items requested*	Action taken

Details of classification

A. Department supplying data		B. Priority	C. Action		D. Cause	
1. Machine shop (in-house)	6.	1. Feedback action required	1. Modify mechanism/ structure	6. Modify parts	1. Planning error	6.
2. Machine shop (outside)	7.	2. As much feedback action as possible	2. Modify action/ function	7.	2. Design error	7.
3. User dept. (engineering)	8.	3. For reference	3. Modify control circuit	8.	3. Fabrication error	8.
4. User dept. (maintenance)	9. Other	4.	4. Modify surface/ heat treatment	9.	4. Spec. change	9. Other
5. User dept. (production)	10.	5.	5. Modify materials	10.	5.	10.

(Note 1) Appended data 1. Operating rate 2. Maintenance history 3. Other (Note 2) Planning staff to complete items marked with *

(Note 3) When two or more items apply, write them in next to each other

From Takahashi, Giichi, *Productive Maintenance Promotion Manual* (Tokyo: Japan Institute of Plant Maintenance, 1975), 550.

	To indicate type of record, circle relevant record name	Date issued:					
		Design			Planning		
		Mgr	Sup'r	Person resp.	Mgr	Sup'r	Person resp.
	Equipment No.:						
	Construction No.:						

Date of action	Action (C)	Cause (D)	Result	Feedback destination (E)	Feedback destination (F)	Feedback details (G)	Person resp.	Remarks

E. Feedback destination (equipment)		F. Feedback destination (standards, references)		G. Type of feedback	
1. Future machine	6.	1. DMS	6	1. Increase process capacity	6. Increase safety
2. Similar machines	7.	2. DAS	7.	2. Increase output capacity	7. Increase operability
3. This machine	8.	3. Drawings	8.	3. Increase reliability	8.
4.	9. Other	4. Specifications	9. Other	4. Increase maintainability	9. Other
5.	10.	5.	10.	5. Increase cost efficiency	10.

Table 6-2. Feedback to Equipment Design and Action Taken

Sample Design Standard

Table 6-3 is a design standard for set screws. It is not merely a copy from the industrial standards (JIS) but a living standard growing out of problematic conditions (Table 6-3, paragraph 1: "Motive for preparing new standard"). The scope of application and the schematic explanations of the design and use are simple and clear enough for anyone to understand.

Using Design Standards and Checklists to Eliminate Errors

Standards will not be used if they are too bulky or if information is hard to find, difficult to understand, or out of date. Incorporate standards in training materials, put them in a form that is easy for engineers to use, and revise them in a timely fashion to reflect new data.

Checklists based on design standards can be used effectively for early verifications (debugging). Prepare checklists based on the standards of the most important items at each stage (design, fabrication, installation, test run, and commissioning). Then use these checklists to detect and correct — as early as possible — out-of-standard parts, design and fabrication errors, and other defects. Delay in correcting these items until the commissioning stage increases the cost and makes it difficult to achieve the aim of commissioning, which is to stabilize equipment operation as quickly as possible.

Checklist for Design Stage

Table 6-4 is a checklist for the design stage. Note that, in addition to general checkpoints for carrying out design work, separate checklists for preventing design errors are also prepared, and an item is included to ensure that these lists have been used.

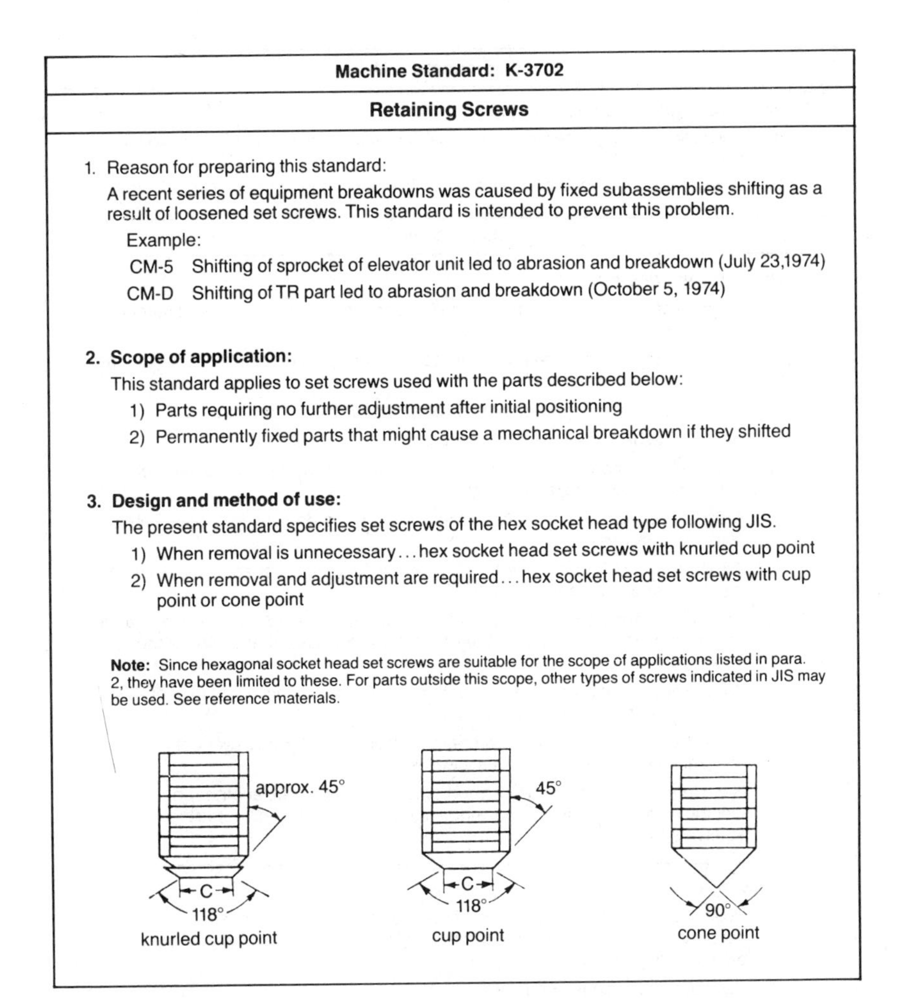

Machine Standard: K-3702

Retaining Screws

1. Reason for preparing this standard:

   A recent series of equipment breakdowns was caused by fixed subassemblies shifting as a result of loosened set screws. This standard is intended to prevent this problem.

   Example:

   CM-5 Shifting of sprocket of elevator unit led to abrasion and breakdown (July 23,1974)

   CM-D Shifting of TR part led to abrasion and breakdown (October 5, 1974)

2. **Scope of application:**

   This standard applies to set screws used with the parts described below:

   1) Parts requiring no further adjustment after initial positioning
   2) Permanently fixed parts that might cause a mechanical breakdown if they shifted

3. **Design and method of use:**

   The present standard specifies set screws of the hex socket head type following JIS.

   1) When removal is unnecessary... hex socket head set screws with knurled cup point
   2) When removal and adjustment are required... hex socket head set screws with cup point or cone point

Note: Since hexagonal socket head set screws are suitable for the scope of applications listed in para. 2, they have been limited to these. For parts outside this scope, other types of screws indicated in JIS may be used. See reference materials.

Table 6-3. Design Standard for Set Screws (Fuji Photo Film)

Acceptance Standard for Equipment Completion

Table 6-5 is a sample standard for lubricating devices. It first asks whether standard parts (from the design standards) have been used and then lists criteria for the most important tasks to be checked at the time of equipment completion, such as installation, pump assembly, piping, and so on.

Design standard: Y-0302
Design Checklist (Preconditions) 1. Tackle the task positively. Use sound technical knowledge and a scientific approach to achieve the best results. 2. Work in active collaboration and cooperation with other departments concerned.
Check Details
I. Planning and design: 1. Do you know the object of the design? (reason, conditions set by originator of design request relevant conditions) 2. Are the design procedures appropriate? (method of execution, completion date, priority schedule planning) 3. Is there satisfactory contact with the originator of the design request? 4. Has the site been thoroughly investigated? 5. Are reference materials adequate? (technical data, introduction of new technology use of existing technology) 6. Are the most suitable and optimal methods and systems being applied? Has complacency been avoided? 7. Are design calculations error-free? (strength, functions, capacity) Have all problems been fully considered? 8. Are maintenance prevention considerations adequate? (Was the maintenance department consulted and did they confirm in advance; will they check the design afterward?) 9. Will the design be cost-efficient? (within budget, operating costs) Is operability good; has safety been considered? 10. Has the optical sensitivity of materials been checked? (Has a request for the photographic characteristics test been issued?) 11. Have related departments been contacted?(maintenance engineering, electrical and instrumental, packing engineering, fabrication departments, safety department)
II. Drawings 1. Have the drawings been reviewed? Are they error-free? (dimensions, number of parts, accuracy, materials, procurement of spare parts, use of checklists to prevent design errors) 2. Has microfilming been considered? 3. Have cost reduction checklists been used? 4. Have the drawings been checked and approved?
III. Purchasing 1. Are specifications of equipment to be purchased satisfactory? (use of standard documents, selection of equipment) 2. Are purchasing arrangements satisfactory? (no mistakes in the arrangements, delivery times, prices, selection of manufacturers) 3. Have vendor's estimates been thoroughly reviewed? (prices, delivery times, details)

Table 6-4. Design Checklist (Fuji Photo Film)

Nippondenso Mechanical Engineering Standard DMS 1-025001 B	
Lubrication Device Standard	**Issued:** August 25, 1969 **4th revision:** March 19, 1977 **__th confirmation:** ________

4.1 Items to be performed by manufacturer

(3) In principle, the parts used in lubricating devices must be those specified in DMS standards.

4.2 Installation

4.2.1 (1) (a) Lubricating devices must be installed or protected so that they cannot be damaged by falling objects, careless material handling, or careless actions of workers.

(b) Lubricating devices must not be installed where they may overheat.

(c) Lubricating devices must be installed where they can be easily adjusted, repaired or replaced.

(d) All lubricating devices must be installed so that they do not interfere with the adjustment or maintenance of plant equipment. They must also not be installed where they may hinder normal work.

(2) Lubricating devices must not be installed in locations where operators will have to reach over rotating main shafts or tools in operation to supply oil or otherwise attend to the devices.

4.2.2 Unless specifically required by their dimensions or function, control devices must be installed between 30 cm and 180 cm from the work floor.

5.1 Installation of pumps

(1) Pumps and associated equipment must be installed in easily-accessible positions for maintenance.

(2) Pumps should be installed on the outside of lubricating reservoirs.

6.3 Piping

6.3.1 Piping joints must be designed and installed for rapid assembly and disassembly using hand tools.

6.3.2 Piping from the end of one lubrication part to the next must not be jointed on the way by welding or any other method. Joints must not be used except when required for length adjustment or assembly. Piping must also be removable without removing any plant equipment parts.

6.3.3 Piping must not be installed where it will interfere with normal operation, adjustment and repair of equipment or with replacement of lubricating devices and cleaning of oil reservoirs.

7.1 Construction of oil reservoirs

7.1.1 Oil reservoirs must be constructed so as to prevent the ingress of water or other foreign particles and to prevent oil leaks and bleeding.

7.1.2 Oil reservoirs must be constructed for easy cleaning and draining.

7.1.3 Oil reservoirs must have an oil supply port. Oil supply ports must be fitted with a strainer and have a suitable cap or cover. Methods must be devised to prevent the cap or cover from being lost.

7.1.4 Oil reservoirs must be fitted with a level gauge positioned so that the oil level can be seen from where the oil is supplied.

8.1 Filters and strainers

8.1.3 (1) Filters and strainers must be fitted with filter media which are easily replaced without stopping equipment.

Table 6-5. Lubrication Device Standard (Nippondenso)

Equipment Standard K-4412

1. **Major dimension check** (compare with dimensions specified on drawings)
   - Radius of rotation
   - Arm
   - Chucking nose
   - Dimensions of bed and position of anchor bolt holes, etc.
2. **Finished precision check** (as per precision specified in relevant drawings)
   - Horizontality of chucking nose (within 5/100)

Arm	Chucking		Unchucking	
	Unloaded	Loaded	Unloaded	Loaded

   - Horizontality of chucking nose (within)

Arm	Chucking		Unchucking	
	Unloaded	Loaded	Unloaded	Loaded

   - Horizontality of auxiliary rollers (within 5/100) • Horizontality — ( )
   - Parallelism • Parallelism — ( )
3. **Assembly condition, check**
4. **Actuation check**

Table 6-6. Designed Function Checksheet (Fuji Photo Film)

Checklists for Installation and Test Run

Table 6-6 is a design capacity checklist to inspect newly installed equipment in loaded and unloaded conditions. Table 6-7 is an example of a data sheet for use during testing, requiring that the dynamic precision of the rollers be measured and recorded. Table 6-8 (*see* p. 309) is a detailed list of reliability and maintainability check items. Great care is taken to ensure that not a single minor defect is overlooked.

Roller Speed Check Sheet **1) Date of measurement: November 20, 1975**

Roller Dia. \ Counter speed		39.6 – 40.1 m/min		70.0 – 70.1 m/min.		(m/min.)	
		Measuring instrument A (m/min)	Measuring instrument B (rpm)	Measuring instrument A (m/min)	Measuring instrument B (rpm)	Measuring instrument A (m/min)	Measuring instrument B (rpm)
1	120	(38.6–38.7)	105–106	(67.5)	185–186		
2	318	39.3	39–40	69.2	69–70		
3	120	—	105–106	—	185–186		
4	120	39.3–39.4	105–106	68.7–68.8	185–186		
5	—	—	—	—	—		
6	80	39.3–39.4	158–159	69.1–69.2	278–279		
7	120	38.9–39.0	105–106	68.5–68.6	185–186		
8	120	38.8–38.9	105–106	69.0–69.1	185–186		
9	120	39.3–39.4	105–106	68.5–68.6	185–186		
10	100	[39.5–39.6]	126–127	69.0–69.1	222–223		
11	120	39.3–39.6	105–106	[69.3–69.4]	185–186		
12	120	(38.6–38.7)	105–106	68.5–68.6	185–186		
13	120	39.0–39.1	105–106	68.7–68.9	185–186		
14	100	39.3–39.4	126–127	69.2–69.3	222–223		
15	120	39.2–39.3	105–106	69.2–69.3	185–186		
16							
17							

( ): Minimum value
[ ]: Maximum value

Table 6-7. Sample Test Run Data Sheet (Fuji Photo Film)

Predicting Problems and Preparing Checklists for Subsequent Stages

Debugging will be incomplete if only standard checklists such as those described above are used. Items unique to a particular machine will not be covered by a standard checklist and may cause problems. Study the details of the checks and the results of actions taken at each stage and identify priority items for checklists to be used at subsequent stages.

COMMISSIONING CONTROL

Commissioning is the stage when salable products can be produced, following installation and test running to detect and correct outstanding errors in the equipment and attain stable operation quickly. (Installation and testing are sometimes included in early commercial operation control and are together called "test run.") After the equipment is debugged and operating correctly, it is handed over to the operation and maintenance departments for normal operation.

The Importance of Control

Prior to commissioning, check after check is performed and every effort made to prevent defects from being carried over into the commissioning stage. Commissioning is the last opportunity to detect and correct design defects that are impossible to predict. Frequent breakdowns at this stage indicate that debugging in the previous stages has been less than thorough.

Debugging at the commissioning stage should only identify problems of product quality stabilization, feeding raw materials, and material handling on the shop floor. At this stage, prepare operation and tooling manuals and routine maintenance standards for lubrication, inspection, routine servicing, and so on. Train the relevant production and maintenance personnel to take over the equipment when it is handed over.

Commissioning Control System

Figure 6-5 (*see* pp. 311-312) is a flow chart for a commissioning control system. In this system the roles of the production, maintenance, and planning (engineering) departments are clearly defined and the activities are carried out on a cooperative basis. Commissioning is most successful when this type of system is used.

Commissioning Initiation and Cancellation

In this system the initiation and cancellation procedures for commissioning control are clearly defined. At the initiation of commissioning control, cancellation criteria such as production performance, stoppage frequency, stoppage loss time, quality

Item	Points of interest and method of checking
1. Are screws fitted with locking aid?	Are locknuts, spring lock washers, and locking compounds in use?
2. Are welds of satisfactory strength?	Is weld overlay adequate? Examine cut welded portion with special care.
3. Are shock-absorbing devices effective?	Is there any shock that will affect parts lifetimes? Do shock absorbers work, and are they controlled?
4. Are parts adequately finished?	Is there any chance of scuffing or defective movement through inadequate finishing? Compare with drawings and modify if necessary.
5. Can parts be replaced?	Give priority to examining areas where deteriorated parts or consumable items must be replaced.
6. Are there any easily-fatigued or damaged parts?	Have any such parts or dangerous parts become obvious during test run?
7. Are any geared belts subject to pitching?	Check during test running.
8. Are positioning methods adequate?	Can positioning be secured accurately through the use of positioning notches, guides, and so on?
9. Are any parts rusting?	Check rusted parts or parts that seem likely to rust. Is surface treatment adequate?
10. Are springs properly assembled?	Are any springs subject to unreasonable strain due to assembly method, compression, or tension?
11. Are arms, brackets, and studs properly attached?	Is any bending or twisting observed during test run? Are these parts securely assembled?
12. Hydraulic cylinder assembly, oil leaks	Is there any oil leakage from hydraulic cylinder? Is assembly method as designated and of adequate strength?
13. Installation and locking of speed controllers	Are speed controllers installed properly? Are speed gauges and locks attached?
14. Roller and bearing replacement	Can rollers and bearings be replaced?
15. Are there any places where tools cannot be used?	Can tools be used in places where adjustments are required? (guides, arm positions, shearing machines, packing machines, etc.) Or are special tools needed?
16. Are covers easy to handle?	Are the safety covers of drive mechanisms and edged parts securely fixed, safe, and easily handled?
17. Is wiring securely fixed?	Is all wiring inside machinery securely fixed, out of contact with moving parts, and properly sheathed?
18. Are all cable connectors properly prevented from loosening?	Are all cable connectors firmly inserted and not loose?
19. Are brush and commutator properly contacted?	Are commutator surfaces, brush contacts and attachment methods satisfactory? Is there any slackness?
20. Are foreign particles being thrown up by gears and belts?	Is any powder or other foreign matter being thrown onto the workpiece from plastic gears, synchronous belts, etc.?
21. Limit switches	Are limit switches installed in easily visible positions? Is there proper contact with the toggles?
22. Are shafts and couplings easily replaced?	Can these be dismantled and assembled without affecting other parts or their accuracy?
23. Replacement of clutches and brakes	As above. Is the wiring securely fixed?

Table 6-8. Maintainability Checklist (Fuji Photo Film)

defect rate, and so on, are specified in advance. Table 6-9 (*see* p. 313) is an example of a commissioning control initiation/cancellation notice.

Commissioning Data Control

This system facilitates accurate recording of data on problems and improvements during the commissioning period using a method similar to the MTBF analysis chart and card method introduced in Chapter 5.

Equipment Handover Documents

When commissioning is completed and the equipment is handed over to the production and maintenance departments, the necessary design materials and data, standards, manuals, procedures, and other materials must be handed over at the same time to ensure that there is no obstacle to operating and maintaining the equipment. Table 6-10 (*see* p. 314) shows an example of an equipment handover document.

PROBLEM PREVENTION METHODOLOGY

As described above, problem prevention activities are intended to eliminate — at the equipment planning, design, fabrication, installation, and commissioning stages — many problems that typically occur during and after commissioning. Their most important goal is to predict the possibility of trouble at the commissioning stage and after.

This is accomplished by using standard checklists at each stage, by observing the results of preventive measures taken, and by considering items unique to particular pieces of equipment. At each stage, the high-priority measures and items to be checked at the following stages must be identified.

This kind of preventive activity is seldom pursued in practice, however, so the number of problems at the startup stage

and after does not decrease. The following approach to problem prevention (early equipment management) can be used to help avoid this.

Forming the Project Team

Equipment design is usually considered the job of the engineering and design departments and is often left entirely up to them. To reduce the number of problems occurring during and after the commissioning stage, however, the people who will operate and maintain the equipment should participate from the design stage onward. Production and maintenance departments must join forces with the design department to form a project team centered in the engineering department.

The production and maintenance departments can examine the problems at each stage from their different perspectives and experience. For example, production can advise the team on quality, productivity, operability, safety, and environmental issues; the maintenance department can advise on reliability, maintainability, and energy-saving. This leads to more accurate and thorough prediction of problems and to preventive measures that the engineering department could not provide alone.

Using Problem Prevention Control Charts

The project team needs appropriate tools to carry out early equipment management reliably and effectively. The step-by-step problem prevention control chart shown in Figure 6-6 is an example. Using this chart, the team predicts or anticipates potential problems at each stage of early equipment management, plans preventive measures in advance, and checks the results after the measures are carried out.

Planning Stage

First, the project team members meet during the planning stage. Using their past experience and knowledge of similar

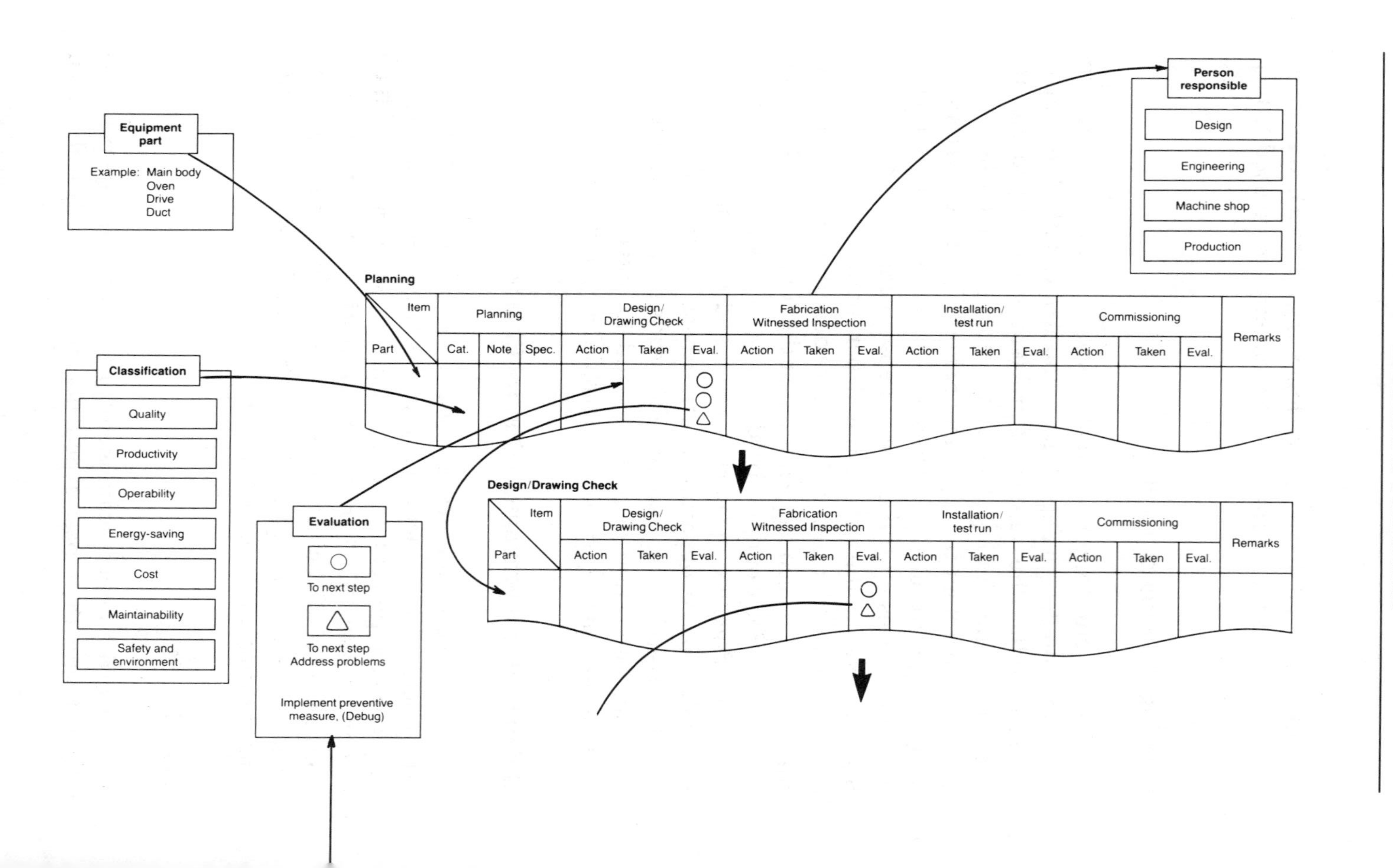
Equipment part
Example: Main body
Oven
Drive
Duct
Planning
Item
Part
Planning
Cat.
Note
Spec.
Design/ Drawing Check
Action
Taken
Eval.
Fabrication Witnessed Inspection
Action
Taken
Eval.
Installation/ test run
Action
Taken
Eval.
Commissioning
Action
Taken
Eval.
Remarks
Person responsible
Design
Engineering
Machine shop
Production
Classification
Quality
Productivity
Operability
Energy-saving
Cost
Maintainability
Safety and environment
Design/Drawing Check
Item
Part
Design/ Drawing Check
Action
Taken
Eval.
Fabrication Witnessed Inspection
Action
Taken
Eval.
Installation/ test run
Action
Taken
Eval.
Commissioning
Action
Taken
Eval.
Remarks
Evaluation
To next step
To next step Address problems
Implement preventive measure, (Debug)

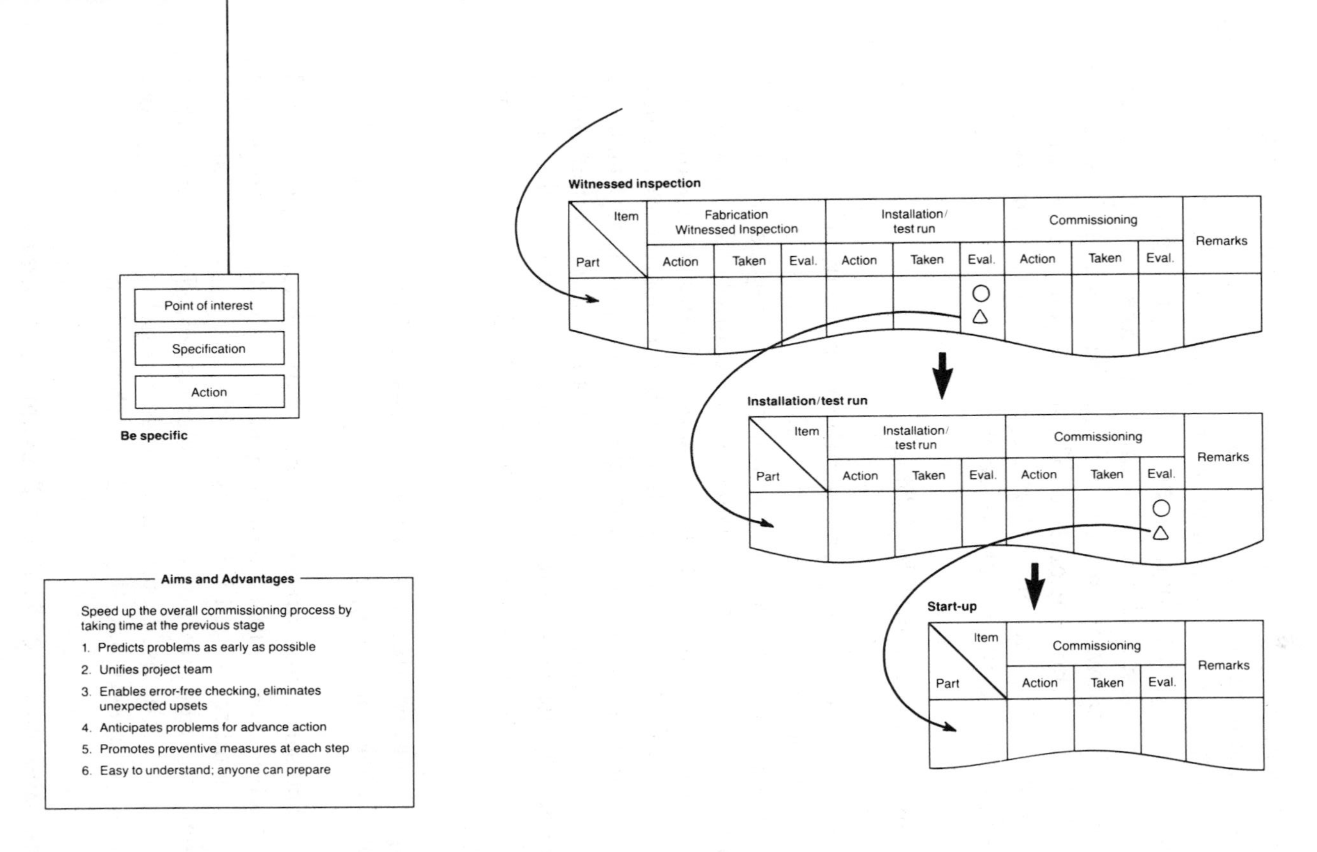

Figure 6-6. Step-by-Step Problem Prevention Control Chart

equipment, they list the issues and specifications that must be clarified at the planning stage. The team uses the basic specifications prepared by the planning or engineering department as their starting point. Then, for each issue they identify points that must be discussed, checked, or otherwise followed up, from the design and drawing-check stage onward. These points are recorded across the control chart. The action that must be taken is recorded in the "action" column at each stage, and individuals are assigned responsibility for each item.

Design and Drawing-Check Stage

At the design and drawing-check stage, the project team members meet to carry out and follow up on the results of the actions taken and recorded at the planning stage. Items with satisfactory results are indicated on the control chart by a circle, while those with unsatisfactory results are indicated by a triangle.

A blank design control chart is then prepared for the next stage. Items indicated by a triangle and new problems that arose at this stage are recorded in the "problems" columns. Items to be examined, checked, or otherwise followed up at the fabrication, witnessed inspection, and subsequent stages are recorded on the chart.

Equipment Fabrication, Witnessed Inspection, and Subsequent Stages

The same procedure is repeated at the fabrication, witnessed inspection, and subsequent stages. At each stage, the project team members meet to discuss, check, and follow up on items identified at the previous stage. Problems are predicted, preventive measures are devised, staff members responsible for carrying them out are assigned, and the results are checked and confirmed.

The Advantages of Problem Prevention Control

Problem prevention control charts are sometimes criticized because they take time to prepare and follow. This is true only in the beginning, however. Once the team members are familiar with them, the charts are both quick and easy to use. Even when considerable time is invested in the earlier stages, problems are reduced or eliminated by the commissioning and production stages and the overall early equipment management process is completed quickly. (The advantages of this method are listed in Figure 6-6.) Figure 6-7 shows an example of the effect of using the control-chart method.

LIFE CYCLE COST THEORY

Life cycle cost (LCC) is the total cost of a piece of equipment or system over its entire lifetime.*

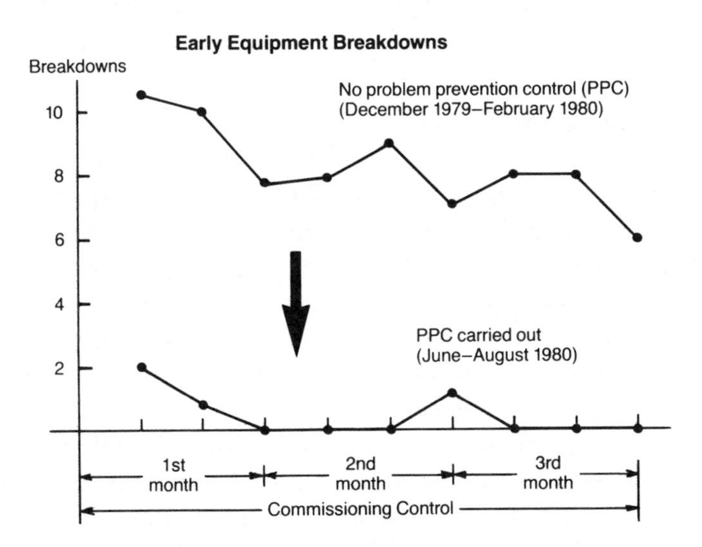

Figure 6-7. Reduction of Early Breakdowns (Comparing Similar Types of Equipment)(Tokai Rubber Industries)

The U.S. Office of Management and the Budget defines it as "the sum of the direct, indirect, recurring, non-recurring, and other related costs of a large-scale system during its period of

---

* The following discussion of basic life cycle costing is based on Benjamin S. Blanchard's *Design and Manage to Life Cycle Cost*, (Forest Grove: M/A Press), 1978.

effectiveness. It is the total of all costs generated or forecast to be generated during the design, development, production, operation, maintenance, and support processes."*

Expressed more simply in terms of production equipment, this means that LCC is the total cost of design and fabrication (the initial or acquisition costs), plus the costs of operation and maintenance (*i.e.*, running or sustaining costs), which can be surprisingly high. LCC can also be viewed as the total cost required for a system to fulfill its mission. *Acquisition costs* are the one-time costs incurred in acquiring the system. *Sustaining costs*, however, occur continuously from the time the system is first used to when it is scrapped, and thus include costs generated over the long term. In a 1970 study of consumer products, for example, the ratio of life cycle cost to original price ranged from 1.9 to 4.8.**

What is Design-to-Cost?

In design-to-cost (DTC), the LCC of a system (equipment) or item of machinery is included as one of the design factors along with the operating precision, speed, volume, machine weight, reliability, maintainability, and other design factors.

In DTC, the cost is not considered a result of the design process but rather one of the system targets. Thus, once the target cost of the whole system has been decided at the preliminary development stage, the amount is apportioned first among the first-level subsystems, then among the second-level subsystems, and so on. In this way, the cost is distributed among all subsystems. The aim is to achieve the target costs using a variety of scientific methods. Figure 6-8 outlines this process. As Figure 6-10 shows, if we consider the life cycle of the equipment or system, 80 percent or more of the LCC is fixed at the conceptual and preliminary system design stages.

---

* U.S. Office of Management and Budget. 1976. *Major System Acquisitions*. Circular #A-109.

** F.M. Gryna, Jr., "Quality Costs: User vs. Manufacturer," *Quality Progress*, June 1970, 10-13.

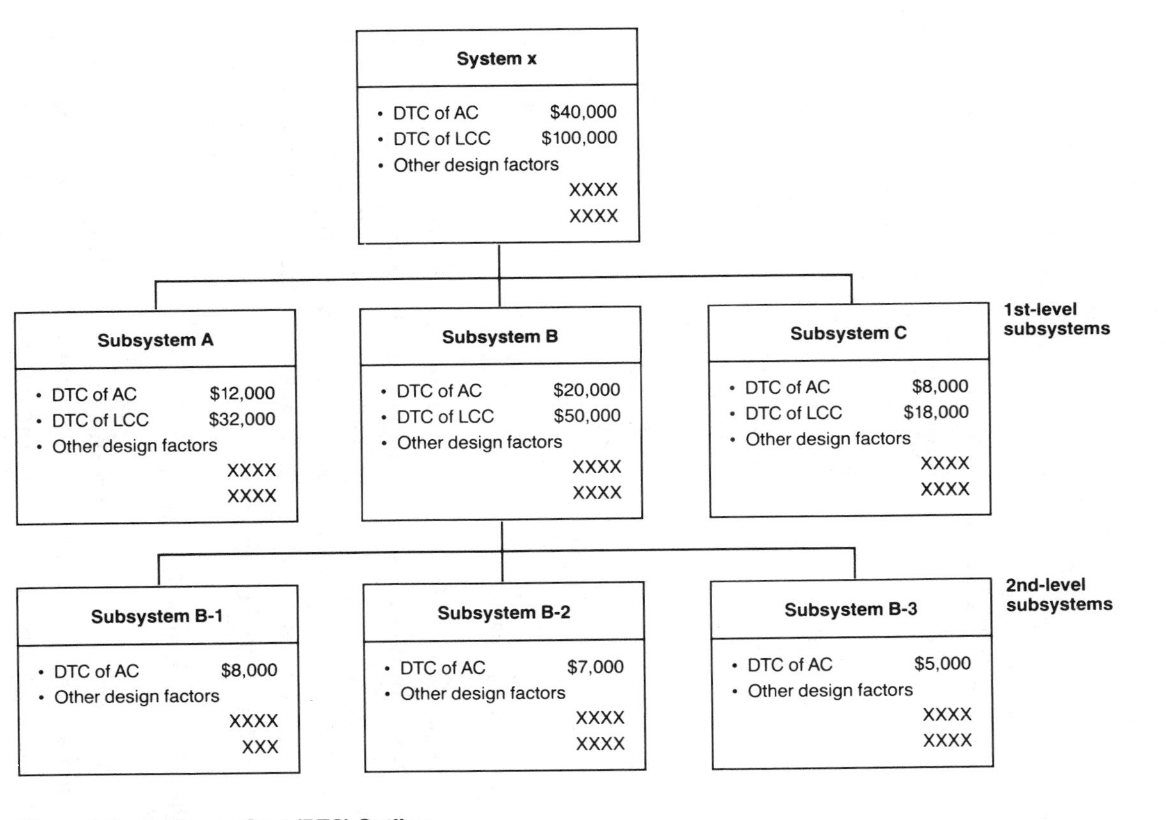

Figure 6-8. Design-to-Cost (DTC) Outline

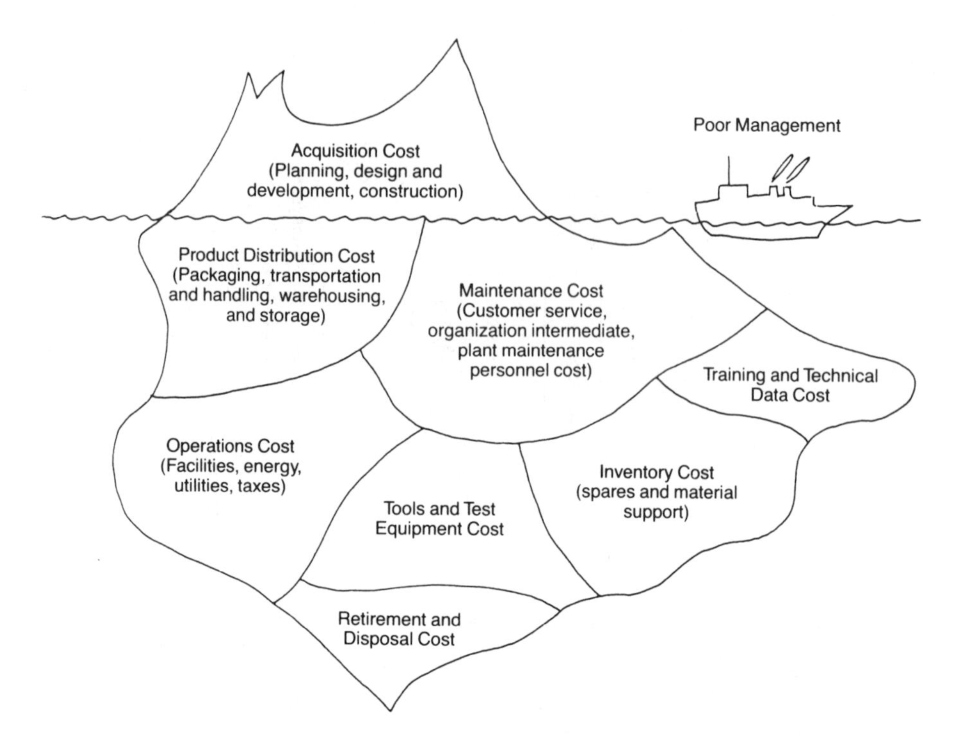

Figure 6-9. Total Cost Visibility

The true LCC is not usually identified, however. As Figure 6-9 shows, only the tip of the cost iceberg is visible. In addition to the initial purchasing costs, which are easily identified, the running costs must also be considered. Serious problems have been caused in the past by failing to give adequate consideration to the costs of running equipment.

What is Life Cycle Costing?

According to Professor Blanchard, life cycle cost analysis is a systematic, analytical approach to selecting the optimal method for utilizing scarce resources from a variety of alternative plans. In life cycle costing, the life cycle cost is treated as a design parameter at the system-development stage and various tradeoffs are pursued to make the life cycle cost of the user's system most economical.

The general procedure for life cycle costing is as follows:

- *Step 1:* Clarify the mission of the subject system.
- *Step 2:* Propose alternative plans capable of fulfilling the mission.
- *Step 3:* Clarify the system's evaluation factors and methods of quantification.
- *Step 4:* Evaluate the plans.
- *Step 5:* Document the analysis results and processes.

The most commonly used formula for comparing design alternatives is

$$\text{cost effectiveness} = \frac{\text{system effectiveness}}{\text{life cycle cost}}$$

System effectiveness is the effect obtained by the introduction of LCC. If LCC is the input, system effectiveness is the output. Usual system outputs include profit, value, utility, and so on.

To pursue an economic LCC, trade-offs at the design stage are needed. The most economical total cost is found by trading off against each other subsystems with mutually contradicting characteristics to achieve optimal balance for the overall system. Here are some of the trade-offs:

- between acquisition costs and maintenance costs
- among acquisition cost items
- among maintenance cost items
- between system effectiveness and LCC
- between acquisition costs and schedule from development to acquisition

An Example of Life Cycle Costing

For a simple example of life cycle costing, assume that a company is considering painting its factory buildings with anti-corrosive paint. Paint from two companies, A and B, is under consideration. Company A's paint would cost $5,000 and last

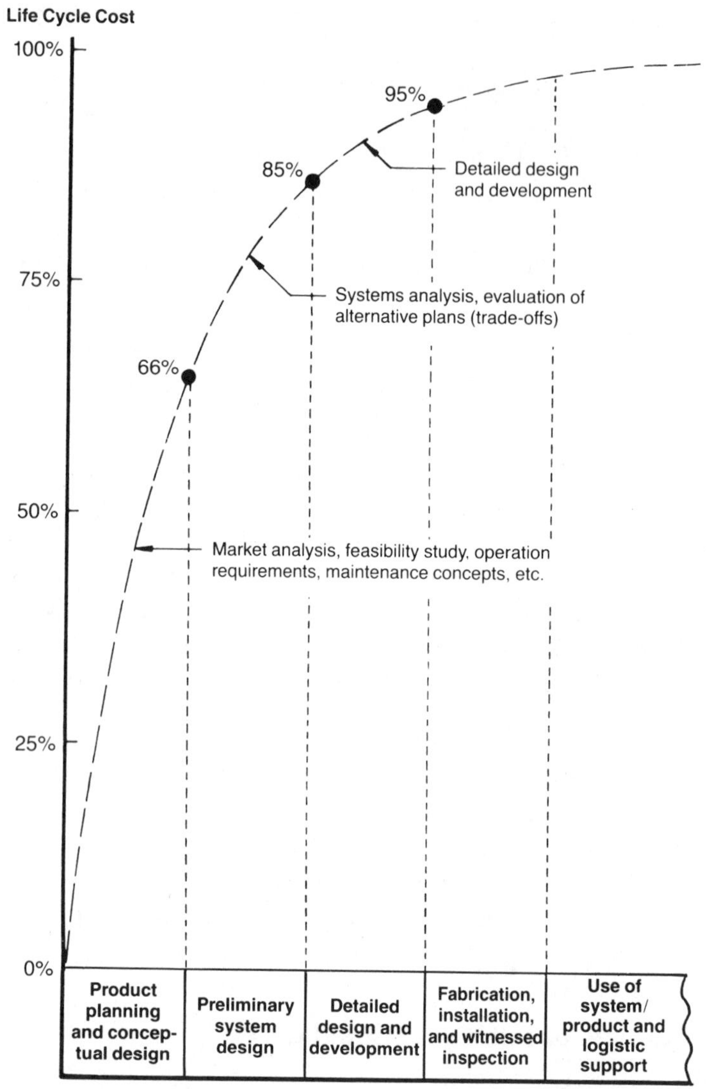

Blanchard, B.S., Design and Manage to Life Cycle Cost (Forest Grove, Ore.: M/A Press, 1978)

Figure 6-10. Business Factors Influencing Life Cycle Cost

three years, while company B's paint would cost $15,000 and last six years. With either paint it would cost $20,000 for labor. Which company's paint is more economical (assuming that interest, changes in price, and technical advances are not considered)?

Comparing the total costs over a period of six years, company A's paint at first appears to be cheaper. Since the paint must be renewed every three years (Figure 6-11), however, an investigation of the total costs (including labor costs) over the whole six-year period shows that company B's paint is more economical.

Comparison of two types of paints

	Cost of paint	Lifespan	Ratio
Paint A	$5,000	3 years	$1,666/year
Paint B	$15,000	6 years	$2,500/year

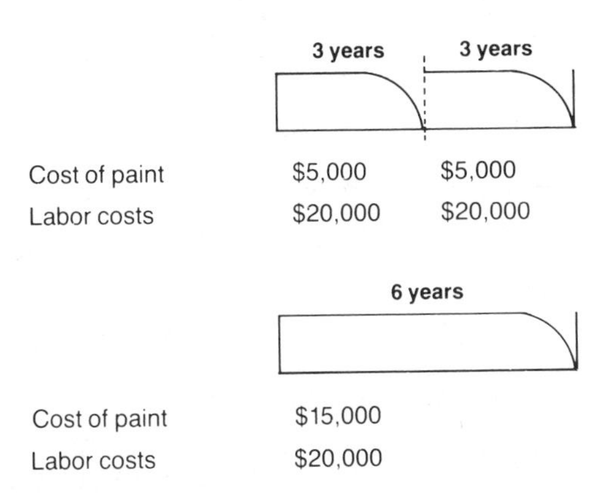

(from Senju Shizuo, Session 1, Third Terotechnology Symposium, 1979)

Figure 6-11. Comparison of Two Types of Paint

This is a simple illustration of the basic theory of LCC. The basic theory remains the same when applied to fabrication and purchase of more complex products, systems, and equipment.

Now consider a comparison of equipment life cycle costs. Procurement estimates obtained for equivalent machines from companies A and B were $100,000 and $70,000, respectively

(Table 6-11). In this case, a simple comparison of the purchase price estimates seems to indicate that company B's machine is more economical, being $30,000 cheaper.

	A	B
Purchasing price of equipment	$100,000	$70,000
Annual sustaining cost	$30,000	$60,000
Life	5 years	
Interest rate	10%	
Annual value method	$56,380	$78,466
Present value method	$213,730	$297,460

(Figures for annual and present values based on economic engineering.)

Table 6-11. A Comparison of Equipment Life Cycle Costs

But what happens if the calculation includes not only the purchase price but also the operating labor costs, energy costs, maintenance costs, and other annual operating and sustaining costs? Assume that these costs amount to $30,000 for company A's equipment and $60,000 for company B's equipment. Then apply concepts of engineering economics to perform an equivalence conversion over 5 years at a 10 percent rate of interest. Comparisons by the annual value and present value methods show that company A's equipment is more economical.

REFERENCES

Blanchard, Benjamin S. *Design and Manage to Life Cycle Cost*. Forest Grove, Ore.: M/A Press, 1978.

Ikeda, A. "Reliability and Maintainability Design Systems and Their Development" (in Japanese). *Plant Engineer* 9 (September 1977): 28.

Imai, W. "Maintainability Design Activities Mobilize All Departments" (in Japanese). *Plant Engineer* 9 (August 1977): 58.

Morimoto, M. "Reliability and Maintainability Design" (in Japanese). *Plant Engineer* 10 (February 1978): 25.

Senju, Shizuo, Tamio Fushimi and Seiichi Fumita. *Profitability Analysis*. Tokyo: Asian Productivity Organization, 1980.

Sugai, E. "The Role of the Equipment Design Department in TPM with Examples of Reliability and Maintainability Design" (in Japanese). *Plant Engineer* 9 (July 1977): 25.

Takahashi, Giichi. *Productive Maintenance Promotion Manual* (in Japanese). Tokyo: Japan Institute of Plant Engineers, 1975.

Takeuchi, S. "The Development of Maintenance Activities Through TMF Analysis" (in Japanese). *Plant Engineer* 11 (March 1979): 37.

7
Maintenance Skill Training

In recent years, the movement toward increased productivity and more cost-effective production has led to larger and more sophisticated equipment operated at higher speeds. The expertise of both production and maintenance personnel is often limited to issues and technologies related to their own areas of responsibility. Under such circumstances, companywide TPM cannot succeed.

TPM REQUIRES STRONG MAINTENANCE SKILLS

To carry out the TPM activities described in earlier chapters a company needs personnel with strong maintenance and equipment-related skills. Operators — production's front-line workers — must become intimately acquainted with their own equipment and develop the practical expertise and skills necessary to maintain as well as operate it. At the same time, maintenance personnel must be willing to learn and use advanced skills and techniques in responding to the range of equipment problems.

In the maintenance skills training program described in this chapter, the conventional "I operate — you fix" philosophy is gradually replaced by the understanding that successful PM activities require everyone's involvement.

RESPONSIBILITIES OF OPERATORS AND MAINTENANCE PERSONNEL

Operators must understand their equipment's structure and functions well enough to operate it properly. Their primary responsibility is to maintain basic equipment conditions through routine inspection and daily cleaning, lubrication, and bolting. They should also be able to perform simple repairs and parts replacement and other autonomous-maintenance functions. Conversely, to ensure successful operator maintenance activities, maintenance personnel must possess skills and knowledge that operators can rely on.

Equipment Operators Are Like Auto Drivers

The relationship between the two groups can be simply understood by comparing equipment operators to automobile drivers and maintenance workers to repair mechanics (Table 7-1). A driver starting his car is like an operator performing an equipment check at the start of the work day. Even before opening the car door, the driver may walk around the car and inspect the exterior, perhaps checking the air pressure in the tires, cleaning the windshield and the headlights, and noting any problems. Before starting, the driver may also take a look at the engine and make a variety of simple checks — the radiator water level, the condition and level of the oil, the fan belt, and so on — and refill or adjust these items as needed. If he finds a leak in the radiator, however, he must take the car to the repair shop.

The smart driver also keeps an eye on the instrument panel while driving, to maintain a safe speed and to make certain the battery is sufficiently charged, the turn signals and brake lights are functioning, and the engine temperature is normal. The driver who never makes any of these checks may overlook the radiator leak and warning lights and end up with a cracked engine block and major repair bill. Although a driver may make minor repairs such as rotating tires or adjusting the V-belt, the specialized knowledge and skill of an auto mechanic is required to repair a radiator leak or perform regular tune-ups.

Category	Operator	Driver	Auto Service Station	Maintenance Personnel	Comments
Daily Checks	Check equipment before startup (maintenance check); visual inspection of temperature, vibration, etc.	Check before driving; inspect car exterior, engine compartment, instrument panel, steering wheel, etc.	Perform checkups requested by owner and on delivery of new vehicles	Machine and instrumentation inspection with measuring tools	
Minor Repairs	Replenish lubricant, check temperature, clean	Add oil, check fan belt, etc.	Repair at driver's request	Data collection and maintenance planning	Autonomous maintenance
Periodic Repairs	Maintenance log	Tire rotation	Tune-up at six months and inspection/testing every year (required by law in Japan)	Planned maintenance; preventive maintenance, maintainability improvements, and repair record-keeping	
Failures	(various - operator must document)	Punctured tire, engine trouble, etc.	Repair	Correct cause of breakdown and resume operation quickly	

Table 7-1. Automobile and Industrial Equipment Maintenance

The Operator's Four Basic Functions

Like the automobile driver, the equipment operator performs four types of simple maintenance work to keep equipment running smoothly. The operator

1. conducts an equipment spot check at startup time, checking the oil level in hydraulic systems and the power current value and looking for unusual vibrations or other abnormalities
2. periodically checks the temperature and speed, and so on, during operation and continues to listen for unusual noise or vibrations
3. scans the instrument panel regularly to check the power-current level and the various other meters and gauges
4. makes certain the equipment is well lubricated by replenishing lubricant whenever it is needed

Finally, like the auto driver who discovers a radiator leak, an operator who notices a change in the equipment's condition that poses a safety or mechanical problem informs the maintenance department that the machine is operating abnormally and requests a thorough examination. The maintenance personnel can then go to work immediately to find the causes of the abnormality.

Optimal Functions for Maintenance Personnel

In working on a problem, the maintenance worker checks the equipment records to determine whether the equipment has had a similar problem before. If so, he checks the previous repair record in order to estimate the man-hours and spare parts required. Naturally, the worker will prefer to carry out the work required in a way that will restore the equipment to operation quickly and minimize any production slowdown.

Although maintenance personnel strive to carry out breakdown maintenance as promptly and efficiently as possible, their duties go beyond treating equipment failures. Maintenance personnel have always been responsible for ensuring the reliable operation of machines and other equipment used by the production department. Therefore, their duties include

- periodic planned maintenance (overhaul)
- periodic vibration and temperature measurements
- estimating optimal overhaul and parts replacement intervals
- planning and selecting the optimal lubricants, material, and machine parts
- correcting equipment design weaknesses
- restoring equipment breakdowns promptly
- providing maintenance education and training for equipment operators
- improving their own maintenance skills and learning new technologies

To ensure the long-term durability and reliability of equipment, maintenance personnel study equipment deterioration patterns. They continuously gather and analyze data on equipment abnormalities that will help prevent breakdowns.

For example, maintenance personnel collect various kinds of data from the equipment operators. They develop methods for measuring and diagnosing equipment conditions. They also calculate the approximate life of equipment so that necessary repairs can be made before it breaks down. Finally, they study and systematically improve or remodel individual machines to increase their maintainability.

MAINTENANCE SKILLS TRAINING: OBJECTIVES AND CURRICULUM*

Maintenance workers handle many types of equipment, but all equipment is made up of certain common parts. The maintenance skills training outlined in this section is organized around

* The Japan Institute of Plant Maintenance conducts seminars and workshops on maintenance skills for maintenance personnel. The maintenance skills training described here is based on the basic course in machine-maintenance skills offered at the JIPM Institute.

these common parts and emphasizes hands-on experience. The program's curriculum is listed in Table 7-2.

(Class hours: 9:00 AM to 5:00 PM)

Unit Topic	Subject	Description (3 days per unit)
1 Bolts and Nuts	Lecture; practice on the shop floor	Opening remarks 1. Orientation 2. How to read drawings 3. Machines and materials 4. Bolts and nuts 5. Material and tightening torque 6. Unit review and comprehension test
2 Keys and Bearings	Lecture; practice on the shop floor	1. Review Unit 1 and answer questions 2. Orientation 3. Fits and tolerances 4. Types of keys 5. Bearings 6. Lubrication 7. Unit review and comprehension test
3 Power Transmissions (gears, belts, and chains)	Lecture; practice on the shop floor	1. Review Unit 2 and answer questions 2. Orientation 3. Gears 4. V-belts 5. Chains 6. Aligning and centering 7. Unit review and comprehension test
4 Hydraulics, Pneumatics, and Sealing	Lecture; practice on the shop floor	1. Review Unit 3 and answer questions 2. Orientation 3. Hydraulics 4. Pneumatics 5. Sealing 6. Cutaway models 7. Unit review and comprehension test 8. Presentation of cutaway models 9. Closing remarks

Note: One unit held per month.

Table 7-2. Curriculum of Basic Machine Maintenance Course

Unit 1: Bolts and Nuts

Typically, the many parts that make up production equipment are separately machined and assembled. Most of these parts are joined by bolts and nuts, however. Each machine uses a large number of bolt and nut fasteners that require considerable time to secure and tighten. Loose bolts or screws often cause excess vibration and breakdowns in rotary machinery, for example. In more serious cases, loose fasteners also cause leaks in sealed equipment and create fire or pollution hazards.

While they are vital factors in equipment, these fasteners often do not provide adequate torque. Unit 1 explains adequate torque for different types of material (*e.g.*, carbon steel or alloy bolts). In the process, trainees also learn to read and prepare technical drawings and tables and become familiar with material symbols. They learn the importance of marking off, how to process machine screws and bolts (*e.g.*, by drilling or tapping them), and reliable tightening methods.

Photo 7-1. Students in Training Course

Photo 7-2. Marking Off

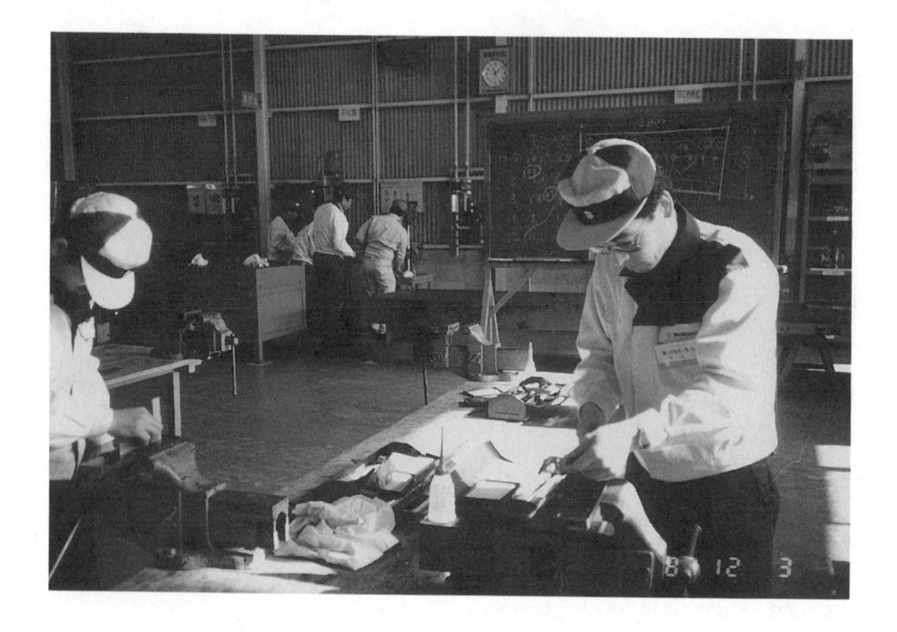

Photo 7-3. Screw Tapping

Unit 2: Keys and Bearings

Keys are important in connecting shafts and parts such as sprockets, couplings, and so on. Keys come in various shapes and sizes according to the load conditions and the structure of the mechanism. Defects in the way shafts and hubs meet or in the way keys fit can cause damage or prevent the rotation of rotary machine parts.

Consequently, in this unit trainees learn (among other things) about the various types of keys, how they should fit, and how to measure the hub sections of spindles and gears to fit them more precisely. They learn, for example, how to file round steel bars and how to correct surface flatness. Once they are able to process flat surfaces using files, they also learn to use driving keys and stud keys together. They discover firsthand how keys with small interference fits cannot fasten spindles and hubs firmly. The trainees also learn to replace keys and are given hands-on practice in various key-fitting methods.

By the end of the unit, trainees also have a comprehensive understanding of bearing types, standards, and characteristics, and a general knowledge of lubrication.

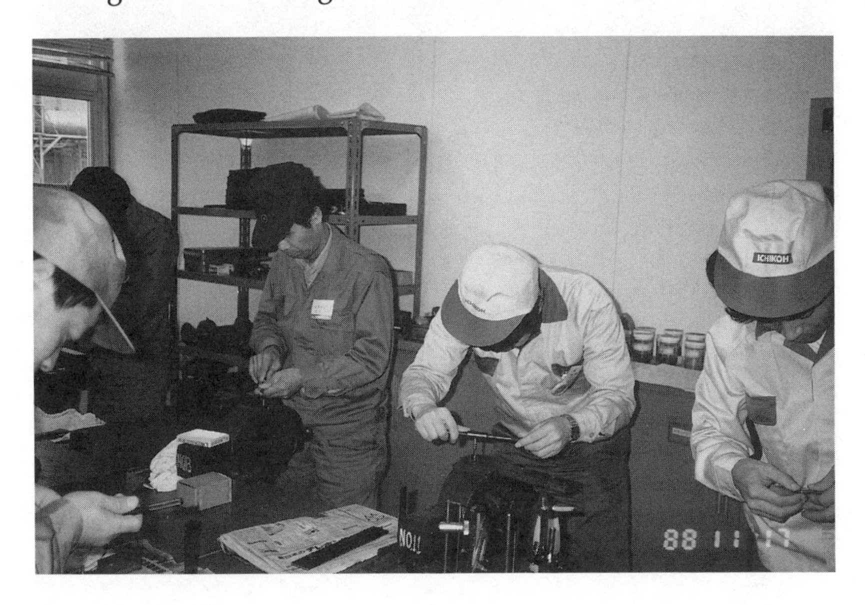

Photo 7-4. Key Fitting

Unit 3: Power Transmissions (Gears, Belts, and Chains)

After an introduction to gear nomenclature, drive chains, and V-belts, the trainees begin working with bearings, using double-row ball bearings. Having already worked with keys and V-pulleys in unit 2, they are ready to assemble a complete device for trial operation. They use various instruments to measure assembly defects and bearing defects during the trial operation. By repeatedly adjusting the amount of lubricant, they learn how lubricant volume affects the operating temperature.

Next, they replace the V-belt with a chain, and repeatedly adjust the tension in the chain to get a feel for the relation between chain tension and operating characteristics such as noise and temperature. Then, they connect a key-adjusted gear and practice correcting the backlash caused by the gear as well as the gear alignment.

Photo 7-5. Unit 3: Drive Chain

Photo 7-6. Unit 3: Gear-Driven Device

Unit 4: Hydraulics, Pneumatics, and Sealing

Hydraulic power based on oil pressure and pneumatic power based on air pressure are drive forces used in many types of industrial equipment. To make the most effective use of hydraulic and pneumatic functions, trainees must first understand the structure and characteristics of the equipment they drive. They begin by studying the basic principles of hydraulics and pneumatics and learning to read the symbols used to explain them. Trainees learn how to prevent leakage of fluid and intrusion of foreign matter through the proper use of various types of gaskets and packing. To fully understand how seals work, the trainees assemble machines that use high-pressure pipes and deal with actual leaks caused by water pressure.

The trainees are also required to make cutaway models. These models display parts or subassemblies cut away from the machine as a unit that relate to a certain mechanical function or operation. The trainees attach these parts to a display board and label their functions, creating a model that can be used later for in-house training.

Photo 7-7. Unit 4: Cutaway Model

IMPLEMENTING THE TRAINING COURSE

Table 7-2 lists the curriculum of the course on basic machine maintenance, organized according to the four course-units described above. Each unit takes three days to complete, and the four units are spread out one per month over a four-month period. Each unit is organized around specific topics and other relevant materials approached from a variety of perspectives to facilitate learning.

The ratio of lectures to practice sessions in the workplace is generally three to seven, and this emphasis on hands-on learning helps trainees acquire practical knowledge and skills. The trainees are required to prepare daily reports describing their work, and at the end of each three-day unit, they are asked to summarize and reflect on what they have learned.

The course instructor reviews and evaluates the reports, administers routine comprehension tests, and returns both with comments. At the end of the course, the instructor evaluates each trainee's overall progress. These evaluations are routed by the company's TPM office to the trainees' supervisors, who add their own comments before the reports are returned.

Before going on to the next unit, trainees spend the intervening weeks applying what they have learned at their workplace and passing along their newly acquired knowledge and skills to their coworkers.

During this "pass-along" instruction and practice period, the trainees are encouraged to use a "one-point lesson" method of teaching, focusing on one simple point at a time to simplify instruction, save time, and increase comprehension. The higher-ranking personnel and supervisors are encouraged to provide opportunities for maintenance workers to apply new skills and information in day-to-day work assignments.

Trainees prepare notes on these practical applications and list any questions raised during the process in their daily reports. Like the daily reports, these additional notes are reviewed and commented upon by the instructor and supervisors. Carrying the course into daily work in this manner makes trainees' maintenance work more meaningful. Questions that are not fully answered in the context of daily work and guidance are addressed when the trainees meet with the instructor at the beginning of the next unit.

This approach is followed through unit 3. In unit 4, trainees take on a new challenge: conducting meetings to present their own cutaway models of machine parts. They also hold theme-based meetings with their coworkers to pass along the information they have received at the course lectures.

After completing the four-month period of course-units, the trainees set goals for themselves and strive to achieve them over a six-month period of goal management. During that time they use what they've learned to develop solutions for specific problems in the workplace. When these activities produce positive results, the workers call goal-attainment meetings to announce the results and help promote overall equipment efficiency. Their activities are monitored by management and reported to the company as a whole at three-month intervals.

MAINTENANCE TRAINING FOR EQUIPMENT OPERATORS

Day after day, operators must keep equipment operating normally, but they rarely understand the equipment they use. Knowing little about how and why their equipment functions as it does, many just shrug their shoulders and say "Look, I run the machines. Someone else has to fix them." Productivity never improves in this situation, no matter how often the maintenance staff repair equipment breakdowns.

The maintenance functions taken over by equipment operators in TPM help to overcome this situation and are not particularly difficult. They include the regular cleaning, lubricating, and bolt-tightening needed to keep equipment running smoothly. Operators are also expected to learn to use their senses in daily equipment inspections. When operators assume responsibility for these simple equipment maintenance functions, maintenance personnel are free to devote more time to equipment diagnosis and maintainability improvement, with the ultimate goal of producing maintenance-free equipment.

The maintenance training program for equipment operators recommended by JIPM includes one week of basic information and skills training at the company's training facility, followed by daily one-on-one instruction on the factory floor by maintenance personnel.

Topics covered in operator maintenance training include handling and maintenance of

- bolts and nuts
- shafts and couplings
- bearings
- gears
- power transmissions, sprockets, V-belts, and chains
- sealing
- lubricants and lubrication

RELATED COURSES

Several other courses help promote thorough maintenance: instructor training, maintenance procedures, electrical wiring and instrumentation, and equipment diagnostic tools.

Instructor Training

This course helps prepare instructors for in-house training programs. Maintenance personnel who have already passed at least the basic maintenance course "sit in" on basic courses in progress to study how they are taught. As part of the training, they are asked to prepare and teach a portion of the curriculum. This experience builds their confidence as teachers, tests their maintenance skills, and provides an opportunity to practice leadership. At the end of a three-day unit, the instructors-course trainees gain experience in giving advice by reviewing their students' daily reports and adding their own comments. They also grade the comprehension tests and are asked to reflect on the effectiveness of their own guidance.

The instructors-course trainees perform similar tasks in each of the units, which helps build their confidence. After completing the course, these freshly trained instructors are encouraged to tailor the basic curricula for in-house courses in their own plants.

Maintenance Procedures Course

Basic maintenance skills can be taught through in-house maintenance training programs. To carry out actual repairs, however, workers need standardized maintenance procedures and reference values for activities such as removing bolts and nuts, opening covers, or removing gears without match marks and reassembling them so the gear teeth mesh correctly.

The maintenance-procedures course at JIPM uses the company's own machine parts as teaching materials and follows the type of curriculum shown in Table 7-3. Photos 7-8, 7-9, and 7-10 were taken during this course.

Unit	Goals	First Day	Second Day	Comments
1 **Dismantling** Dismantling, cleaning, and inspection	• Learn the structure and functions of machinery • Learn to measure deterioration for more effective preventive maintenance	1. Orientation 2. Create a work schedule 3. Create an inspection checklist 4. Inspection (dynamic and static) 5. Dismantling 6. Cleaning 7. Summary	1. Create sketch 2. Create parts dimension table 3. Specifications and standards 4. Measuring and inspecting parts 5. Ordering new parts 6. Summary	• Limited to two groups (3 to 4 persons) • Daily reports required
2 **Modification** Modify and reassemble	• Learn to reassemble machinery • Learn important points in reassembly processes • Learn to make improvements that extend equipment life	1. Parts modification 2. Investigating causes of deterioration 3. Troubleshooting procedures 4. Minor assembly 5. Summary	1. Assembly 2. Improvements for longer life 3. Sealing 4. Lubrication 5. Summary	Instructor adds comments
3 **Test Run** Trial operation and summary	• Acquire hands-on experience in manual repairs and adjustments • Confirm results through dynamic inspections • Learn the daily maintenance procedures and as maintenance planning	1. Inspection while equipment shut down 2. Manual repairs 3. Logging repairs properly 4. Maintenance planning table 5. Trial operation 6. Summary	1. Creating cutaway models 2. Procedures manual and summary	

Table 7-3. Maintenance Course Curriculum

Photo 7-8. Measuring Parts

Photo 7-9. Measuring Parts

Photo 7-10. Cutaway Model of Pump

Electrical Wiring and Instrumentation Course

Although mechanical failures are the most obvious kind of equipment breakdown, failures also occur in the electrical wiring and instrumentation. The causes of these failures are much more difficult to spot and generally require more time to repair.

Sometimes loose electrical connections cause failures; at other times defective circuits are the cause. When the cause lies within an IC chip, repair time can be lengthy because so much time is needed simply to find the defective chip. Maintenance personnel should be taught the basics of control circuits using simulation programs so they can identify the simpler failures, such as loose connections.

Machine Monitoring

The trend toward larger, more sophisticated, and faster machines is producing increasingly complex machine structures that require condition-based maintenance and the use of various new monitoring and diagnostic techniques. Today's maintenance personnel are obliged to become familiar with these new methods in order to detect abnormalities quickly and prevent equipment breakdowns.

THE IMPORTANCE OF IN-HOUSE TRAINING

In this era of intense international competition, every company's survival depends in large part upon the knowledge and skills of its equipment-maintenance staff. Maintenance personnel need to make full use of the latest equipment diagnostic tools while aiming for the goal of zero breakdowns. The goal of maintenance training must be the development of versatile maintenance professionals who are equally able to handle mechanical, electrical, and instrumentation technologies. Consulting firms like JIPM in Japan teach only a part of the knowledge and skills an excellent maintenance professional requires. To be truly successful, companies must expand the scope and quality of their in-house training.

8
TPM Small Group Activities

A unique feature of TPM is its promotional structure of overlapping small groups, integrating organizational and small group improvement activity.

INTEGRATING SMALL GROUP ACTIVITIES INTO THE ORGANIZATION

Japanese-style small group activities began with the quality control circle, introduced in 1962. The American concept of zero defects (ZD), which is an individual rather than a group activity, became popular three years later. NEC, the first Japanese firm to implement it, combined this individual improvement activity with the Japanese-style quality control (QC) circle to form ZD *group* activities.

Later, the Japanese steel industry followed suit with the widely used *JK* (*jishu kanri*, or "autonomous management,") activities. Since then, many other companies have developed their own terminology and procedures for conducting QC circles and ZD groups. In fact, most Japanese companies now promote some form of small group activity, even service industries such as hotels, banking, and insurance.

QC and ZD Groups Contrasted

In spite of differences in terminology and approach procedures, small groups can be divided into two broad categories, one originating in the early QC circle, the other in the ZD movement (Figure 8-1). These two groups are distinctive in several ways.

QC circles began as study groups to teach quality control techniques to shop-floor supervisors and evolved into problem-solving small groups for a larger segment of the worker population. Circles are organized by subject or theme to deal with specific problems within the larger total quality control (TQC) program. Formed by workers, they are independent of the existing organizational structure. Participation is voluntary. In terms of organizational theory, they are *informal organizations*.

ZD groups, on the other hand, began in the United States at Martin Marietta as a means of involving all employees, individually, in solving the problem of delayed delivery. The Japanese imported the concept and incorporated it in their small group

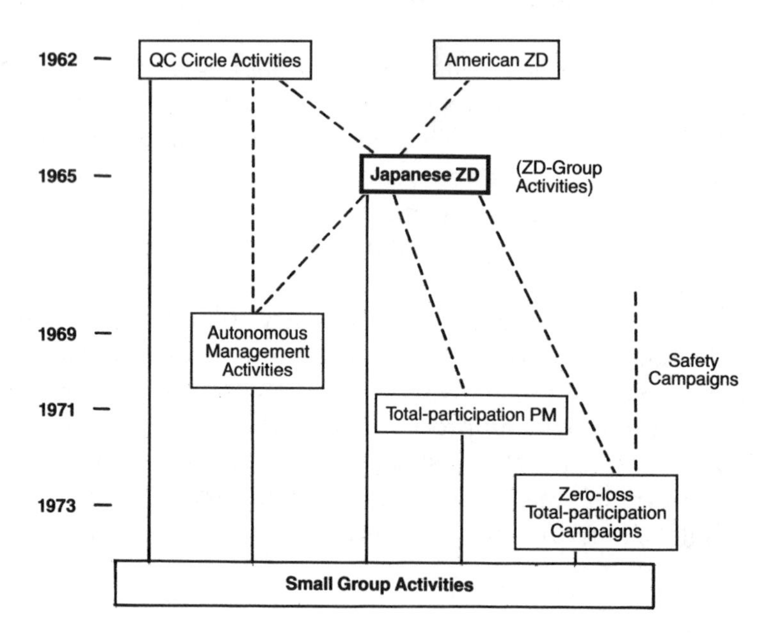

Figure 8-1. Relationships Among Typical Japanese Small Group Activities

activities. Japanese ZD groups participate in management-based activities to solve problems of concern to the company and to work toward company goals.

Formal Versus Informal Organizations

Organizational theory considers Japanese ZD groups *formal organizations* because they are conducted within the existing organizational structure. For example, in the independent QC circles, leaders are typically selected by circle members. In the management-based ZD groups, on the other hand, shop-floor supervisors most often assume leadership roles. Of course, when a supervisor has too many subordinates, subgroups (ranging from 5 to 10 members) must be formed and additional leaders chosen.

Typically, independent QC circle activities are conducted outside of normal working hours, during "free time" (*e.g.*, breaks, after work hours, and weekends and holidays). Because circle activities are voluntary, employees at most Japanese companies do not receive overtime compensation for participation in them. By contrast, the formal ZD groups can meet during work hours under the supervisor's direction as well as during free time, and some companies pay overtime compensation for these outside activities.

Differences in Goals

The improvement themes selected and the goals set also reflect the differences between QC circle and ZD group activities. QC circles are formed around specific themes and goals are set within each theme. Once the goals are achieved, the QC circles are reorganized around new themes. Ideally, themes are selected independently of annual management goals. This is made possible by the informal, voluntary nature of the circles. Although companies appear to respect circle autonomy and allow the circles to choose their own themes, management increasingly encourages TQC activities as part of companywide improvement activities and promotes themes that support the achievement of annual goals.

ZD groups, on the other hand, *must* choose goals consistent with the annual company goals because, ultimately, ZD is aimed at the elimination of defects and promotes the attainment of all related goals. Group members independently discuss and set subgoals such as lower costs, shorter deadlines, and the introduction of new methods.

Although QC circles and ZD small groups differ organizationally, they have often merged and influenced each other, and their distinctive features have become blurred. Many corporations have used both types of groups to develop their own unique systems.

TPM Small Groups Based on the ZD Model

JIPM promotes use of the "autonomous small group" advocated by Professor Emeritus Kunio Odaka of Tokyo University. According to Odaka, Japanese small group activities have flourished, even though their position within the organizational structure has remained ambiguous. He argues that small groups should be integrated into the corporate structure so that their activities can complement and enhance other organizational activities.*

Accordingly, TPM small group activities are based on the ZD model and built into the organizational framework. Small groups function at every level and across divisions to accomplish company objectives. For example, TPM promotes autonomous maintenance by operators through small group activity. In TPM, the typically management-directed activities of equipment cleaning, lubrication, bolting, inspection, and so on, are performed autonomously as small group activities.

* Kunio Odaka has written a number of books on Japanese management, including *Toward Industrial Democracy: Management and Workers in Modern Japan*, published by Harvard University Press in 1975, and *Japanese Management: A Forward-Looking Analysis*, published by Asian Productivity Organization in 1986.

Management of TPM Groups

During the TPM implementation stage, the time spent by small groups on various activities is carefully monitored. Activities are categorized and recorded as maintenance activity, education and training, and meetings, for example. Documenting how small group time is spent allows companies to compensate their employees properly. For example, during the early stages of autonomous maintenance, much time is spent on maintenance and education and training; later on, more time is spent in meetings. Operators should be paid for overtime when they perform maintenance after work hours; employees attending training programs after work hours should also receive education compensation; and if a certain number of hours per month is allotted for meetings, then those exceeding the limit should be held after regular work hours and compensated.

By the time factory workers are able to conduct the autonomous maintenance general inspection (step 4), they can enjoy a real sense of accomplishment when, for example, their efforts reduce breakdowns by as much as 80 percent, increase productivity, and make work easier.

This sense of accomplishment naturally enhances morale and motivation and finds expression in longer and more frequent meetings as well as a greater number of improvement suggestions from workers. Moreover, when maintenance personnel disassemble equipment on weekends for servicing, operators often want to participate in order to learn. By this time, overtime compensation may no longer be an issue.

To promote better-trained and capable operators, managerial staff should lead group activities through step 6 (setting workplace management and housekeeping standards). Workers should be able to carry out autonomous maintenance independently from step 7 on.

Group Goals Coincide with Company Goals

Why do we advocate integrating TPM small group activities into an organizational structure? Here is another perspective:

What does a small group do? According to authors Hirota and Ueda* the small group "promotes itself and satisfies company goals as well as individual employee needs through concrete activities."

Teams called "circles" or "groups" set goals compatible with the larger goals of the company and achieve them through group cooperation or teamwork. This enhances company business results and promotes activities that satisfy both individual employee needs (self-satisfaction, success, motivation) and the needs of the organization. TPM small group activities are representative of this type.

High Morale = High Profits

Behavioral scientist Rensis Likert compared companies and factories that had high productivity with those that had low productivity.** He studied the impact of different management policies and levels of employee consciousness and behavior on productivity.

Likert discovered that the high-producing companies strive to improve product variables (business factors such as profits and sales) as well as intermediate variables (namely, human resources, which serve as intermediaries for the business results) (Figure 8-2). These companies attempt to improve both business results and working conditions. Low-producing companies and factories, on the other hand, ignore the human factor and focus solely on product variables. Likert calls the former "participative" and the latter "authoritarian" management.

Likert argues that participative management is ideal because it encourages confidence among employees and promotes consistently high productivity. Authoritarian management, on the other hand, encourages submission based on fear among

* K. Hirota and T. Veda, *Small Group Activities: Theory and Reality* (in Japanese) (Tokyo: Japan Labor Research Group, 1975).

** Rensis Likert, *New Patterns of Management* (New York: McGraw-Hill Book Co., 1961).

employees. Consequently, even if higher productivity can be achieved for a short time, low employee morale will eventually lead to a decline in productivity.

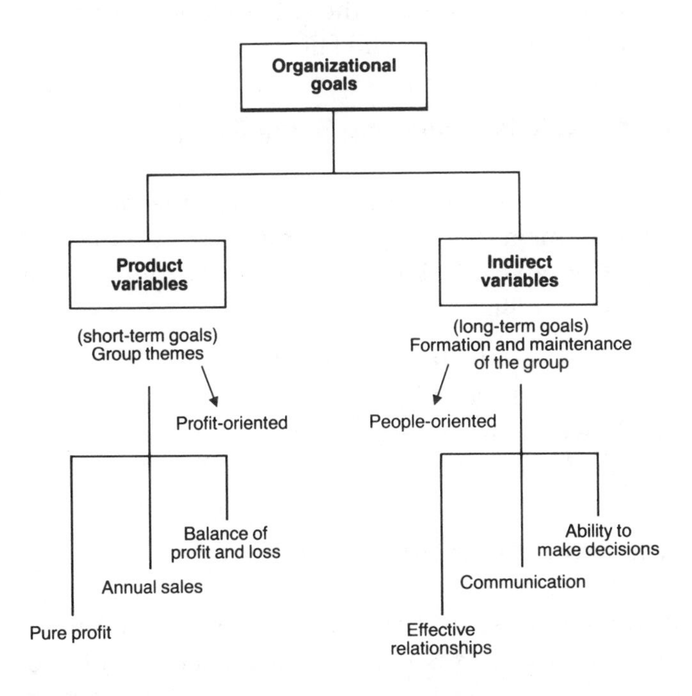

Figure 8-2. Organizational Goals

PROMOTING SMALL GROUP ACTIVITIES

Small group activities in the factory should be based on participative management such as that advocated by Likert. Small group goals should be the same as company goals — to improve productivity and working conditions.

Workers Are the Leading Players

In all small group activities workers are the leading players. In other words, those who do the work take responsibility for it rather than simply following orders and doing enough to earn their pay. Each worker manages his or her own work autonomously, striving for better and better results.

Simply requiring operators to lubricate, clean, tighten, and inspect their machines as part of TPM will have little effect unless operators believe that they are in fact responsible for looking after their own machines. Without the workers' direct support, autonomous maintenance will fail.

Managers' Role in Promoting Small Groups

While workers themselves must play a leading role, at the same time managers must play a large part by working to keep employees motivated. Four important factors in motivating small group activities are

- recognizing the importance of the work
- setting and achieving goals
- acting on workers' suggestions
- rewarding workers' efforts

1. Recognizing the Importance of the Work

A worker must understand the importance of his or her work in order to take responsibility for it and want to do it well. The worker's supervisors and colleagues must also see its importance, otherwise the worker is not going to believe it.

2. Setting and Achieving Goals

Working without a goal is like running a marathon without a finish line — there is no motivation to persevere. Goal management (setting goals and promoting their achievement) can be an effective way to motivate people.

The first step in setting goals is to choose a theme and target items. The second step is to set goal values, and the third is to set the date by which the goals are to be achieved. Managers should take care in guiding small groups so workers are able to select goals directly linked to the annual company goals.

3. Acting on Suggestions

The trend in the number of suggestions proposed by individual workers has been seen as a measure of how enthusiastically small group activities are carried out. *Group* suggestions, based on the creativity and ingenuity of all the members of a small group, are on the increase in Japan. The potential range of suggestions for improving equipment efficiency is virtually limitless. The topic of reducing idling and minor stoppages is particularly well suited to small groups, for example.

If good results are produced when a suggestion is adopted and implemented, the individuals proposing it experience a gratifying sense of achievement. Managers must guide and assist small groups so that members have the opportunity to experience that sense of success. For example, many shop-floor workers have difficulty expressing their thoughts in writing. In many cases, however, the number of suggestions increased dramatically when managers and supervisors took the time to help workers by writing up their verbal suggestions. In plants where small group activities have taken off, every worker is able to contribute at least one suggestion per month.

4. Rewarding Workers' Efforts

Awards satisfy people's desire for recognition and are usually given for achieving goals and for successful suggestions. While these accomplishments alone are satisfying to individuals, the awards have further significance in the context of small group activities. They demonstrate that managers recognize these achievements in a concrete fashion. Therefore, while awards should certainly be monetary, more important than the amount is the opportunity for managers to acknowledge and express their appreciation for workers' achievements.

Leadership in Small Groups

Leadership qualities are extremely important for both small group leaders and managers. Whether small or large, a group is not simply a collection of individuals. It is formed when two or more people come together to achieve an objective, and when cooperation, trust, affection, and other psychological relationships develop through communication among the members.

Broadly speaking, a group has two functions:

- group maintenance: maintaining the existence of the group
- problem solving: attempting to achieve the group's goals

Group maintenance can be seen as the group's social function and problem-solving as its results-oriented or working function.

Leadership within a group consists of helping to set and move toward goals, improving the quality of interaction among group members, increasing the cohesion of the group, utilizing group resources, and so on. In other words, leadership promotes and aids the functions of the group, that is, the group-maintenance and problem-solving functions.

THE ROLE OF TOP MANAGEMENT

The keys to success in all small group activities lie in three conditions: motivation, ability, and a favorable work environment. Management is responsible for actively promoting these three conditions.

Of these three keys, motivation and ability are the workers' responsibility, but the creation of a favorable work environment is outside their control. This environment has both physical and psychological components that must be satisfied.

Figure 8-3 shows the division of these three conditions into two subgroups: human and environmental problems.

Developing Able, Self-Motivated Personnel

Management's first responsibility is to provide the special training necessary to develop a workforce of capable, motivated, and truly autonomous workers. Human education, which explores "what human beings are," as well as technical training in maintenance and operational techniques must be provided.

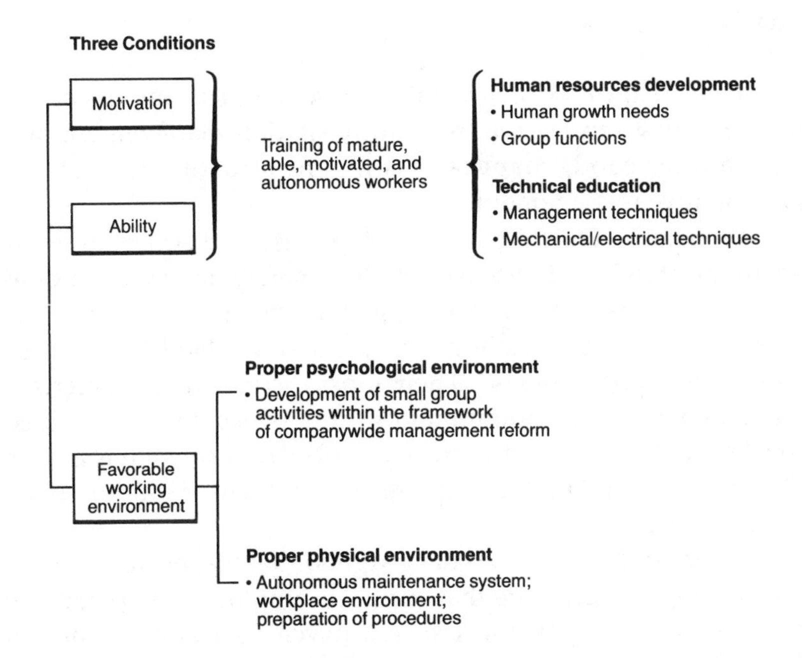

Figure 8-3. Key to Success of Small Group Activities

Education is the source of motivation because it enables people to understand themselves by exploring areas such as human drives and motivation and group dynamics, for example. Unfortunately, managers often have surprisingly little interest in such subjects. However, education essential to the development of mature individuals *begins* with a reevaluation of the self and what it means to be human.

Creating a Favorable Work Environment

Management's second responsibility is to provide a favorable work environment by eliminating the psychological and physical obstacles to worker autonomy in the environment.

The Psychological Environment

The creation of a favorable psychological environment requires, first, an escape from authoritarian management systems and, second, changes in company structure to promote participative management.

William G. Ouichi compared Japanese and American management styles and discovered that management techniques similar to those of the Japanese had been successfully implemented in many leading U.S. companies. He labeled these companies type-Z firms.* Their most distinctive characteristic was a management commitment to employees, which provided the foundation for mutual trust, concern, and egalitarianism. Ouichi concluded that the Japanese do not have a monopoly on such practices.

Likert and Ouichi both have argued that respect for workers and a company structure that supports employees help develop autonomous workers and create a psychological environment that encourages small group activities.

The Physical Environment

The work environment is further enhanced when management establishes certain physical conditions — by creating autonomous-maintenance systems that encourage small group activity, by improving the factory environment, by establishing and adhering to appropriate standards, and so on (Table 8-1). For example, workers' enthusiasm for small group activities may flag if they have no suitable place to hold meetings. Companies where top management is enthusiastic about small group activities often build a lounge in the factory for use as a meeting room.

Autonomous maintenance and (in the maintenance department) preventive maintenance, maintainability improvement, and other types of planned maintenance will improve the equipment and the workplace environment beyond recognition. Seeing

---

* William G. Ouichi, *Theory Z: How American Business Can Meet the Japanese Challenge* (New York: Avon Books, 1981).

their physical environment change dramatically through their own participation in improvement activities gives employees a sense of achievement and further boosts small group activities. Through this experience of re-creating the physical environment, individuals develop and mature and learn to work independently.

Objective	Focus
Design organization for autonomous management:	• Focus on design of responsibilities (fulfilling and enlarging their scope) • Formation
Create a supportive environment in the workplace:	• Workplace organization and housekeeping standards (5 S's) • Proper preparation of equipment, materials, resources, etc • Supportive physical environment (meeting-rooms set aside for activities)
Prepare and adhere to standards:	• Everyone participates in setting and revising operating standards related to their own work, including technical standards, standard procedures, etc.
Establish autonomous management systems for all aspects of the enterprise:	• All systems designed to be managed through autonomous small groups: sales, production schedules, costs, delivery times, safety, personnel matters, remuneration, and work methods • Suggestion and award system revised to be consistent with small group activity

Table 8-1. *Opportunity*: Creating a Favorable Work Environment

EVALUATING THE PROGRESS OF SMALL GROUPS

If small group goals are the same as company goals, we can evaluate their progress by measuring the degree to which group activities contribute to the achievement of company goals. Progress in small group activities can be broken down into four stages (Table 8-2):

Stage 1: Self-development. At first, group members must master techniques; their motivation increases as they recognize the importance of each individual.

Stage 2: Improvement activities. Group improvement activities are proposed and implemented, leading to a sense of accomplishment.

	Stage 1 Self-Development	Stage 2 Improvement Activity	Stage 3 Problem-Solving	Stage 4 Autonomous Management
Summary	Studying techniques and becoming self-motivated	Proposing improvements through group activities and experiencing the satisfaction of achieving targets	Targeting problems that concern the organization and solving them through group activity	Consistently matching small group targets with those of the organization and managing work autonomously
Main Features	• Targets of interest to group members, but often unrelated to those of the organization • Group activities considered separate from work • Groups left alone by top management • Group activities struggle to take off	• Small group targets and organizational targets do not always match; improvement activity is central • Relation between group activity and work not well understood by top management • Leadership and teamwork not always effective	• Small group targets and organization targets match • Group activities are accepted as part of work, but some top managers do not understand them properly • Effective leadership and teamwork	• Small group targets are high and match those of the organization • Group activities are considered part of work, and top management provides active guidance • Superior leadership and teamwork

Table 8-2. Four Stages in Promoting Small Group Activities

Stage 3: Problem-solving. At this stage, small group goals that complement company goals may be selected and the group becomes actively involved in problem-solving.

Stage 4: Autonomous management. The group selects high-level goals consistent with corporate policy and manages its work independently.

The activities in stages 1 through 3 are not inconsistent with a traditional organization based on order and control, with newly implemented small group activities taking place on the shop floor. During stage 4, however, new human resource-oriented organizations are based on the self-managed small group model and are highly motivated to achieve company goals. Thus, during the final stage, true participative management is established. This is the goal of TPM small group activities.

REFERENCES

Hirota, K., and T. Ueda. *Small Group Activities: Theory and Practice*. Tokyo: Japan Labor Research Group, 1975.

Lickert, Rensis. *New Patterns of Management*. New York: McGraw-Hill Book Co., 1961.

Ouichi, William G. *Theory Z: How American Business Can Meet the Japanese Challenge*. New York: Avon Books, 1981.

Nakajima, Seiichi. *New Developments in ZD*. Tokyo: Japan Management Association, 1978.

9
Measuring TPM Effectiveness

WHY MEASURE EFFECTIVENESS?

To integrate TPM programs more effectively throughout the company or plant, the current problems, the potential for their solution, and the benefits to be gained must be clarified at the company and departmental levels. This demands measuring techniques that can isolate the problems and potential for improvement in each department at any time. TPM effectiveness is measured for two reasons: to help establish priorities for improvement projects and to accurately and fairly reflect their results. Effectiveness measurement reveals the fruits of our daily efforts, isolates points we must focus on, and helps us plan countermeasures.

To implement TPM effectively, we must know which areas in the plant are experiencing problems and what those problems are. This requires indices that show — accurately and continuously — where improvement is currently needed and what kind of results can be expected. Such indices focus improvement activities by pinpointing their most important aspects. They facilitate prompt identification and response to change and more accurate judgments, and they help promote more efficient TPM activities. The results of TPM activities are also measured using indices that show — accurately and fairly — the relative effectiveness of activities and improvement measures in the different plants or divisions.

Close monitoring at all levels helps maintain and improve results and promotes the development of more effective countermeasures (where no positive results are being produced). It also helps us understand and prevent sudden drops in effectiveness.

MEASURING EQUIPMENT EFFECTIVENESS

As explained in Chapter 2, overall equipment effectiveness has three factors: availability (operating rate), performance rate, and quality rate.

Overall Equipment Effectiveness

Overall equipment effectiveness is calculated from the formula

$$\text{Overall equipment effectiveness} = \text{availability} \times \text{performance rate} \times \text{quality rate}$$

This is extremely useful as an overall indicator of factory or equipment performance.

Availability (Operating Rate)

$$\text{Availability} = \frac{\text{loading time} - \text{downtime}}{\text{loading time}} \times 100$$

(Downtime includes time for setup, adjustments, changing tools, breakdowns, and other stoppages.) Operating rates can also be calculated for individual types of stoppage using the above formula.

Performance Rate

$$\text{Performance rate} = \text{net operating rate} \times \text{operating speed rate}$$

$$= \frac{\text{output} \times \text{actual cycle time}}{\text{loading time} - \text{downtime}} \times \frac{\text{ideal cycle time}}{\text{actual cycle time}}$$

Quality Rate

$$\text{Quality rate} = \frac{\text{input} - (\text{quality defects} + \text{startup defects} + \text{rework})}{\text{input}}$$

Unit Setup Time

$$\text{Unit setup time} = \frac{\text{total setup time}}{\text{total setup operations}}$$

Reliability and Maintainability Indices

The following measurements are often used to classify and manage breakdowns as priority items:

Reliability Indices

$$\text{MTBF} = \frac{\text{total stoppages}}{\text{loading time}} \times 100$$

$$\text{Failure frequency rate} = \frac{\text{total stoppages}}{\text{loading time}}$$

Maintainability Indicators

$$\text{MTTR (mean time to repair)} = \frac{\text{total stoppage time}}{\text{total stoppages}}$$

$$\text{Failure rate} = \frac{\text{total stoppage time}}{\text{loading time}}$$

Measuring the Efficiency of Maintenance Activities

Indices for measuring the efficiency of the maintenance department differ depending on the type of maintenance work being done and how it is organized. They must address the following general questions, however:

- To what extent is the work proceeding according to plan?

- To what extent is the work helping to raise operating rates and product quality?
- Is the work being done using the most economical and effective methods?

BM (Breakdown Maintenance) Rate

$$\text{BM rate} = \frac{\text{BM jobs}}{\text{total maintenance jobs}} \times 100$$

(Total maintenance jobs = total work done on sporadic breakdowns (BM), maintainability improvement (MI), and planned (preventive) maintenance (PM).)

BM Manhour Rate

$$\text{BM manhour rate} = \frac{\text{BM manhours}}{\text{total maintenance manhours}}$$

PM Achievement Rate

$$\text{PM achievement rate} = \frac{\text{total PM jobs performed}}{\text{total PM jobs planned}}$$

MI (Maintainability Improvement Jobs Trend)

Is the number of MI jobs increasing? What is the trend?

C_p Trend

How did the C_p value change after the completion of work?

Measurement Indices Related to PQCDSM

Table 9-1 shows the various indices related to PQCDSM control items (productivity, quality, cost, delivery, industrial hygiene and safety, morale).

<table>
<tr><th colspan="2">PQCDSM Indices</th></tr>
<tr><td rowspan="3">P = Productivity
Broad measures:
• Labor productivity
• Value added per person
• Overall equipment effectiveness
Specific measures:
• Availability (operating rate)
• Performance rate
• Number of breakdown maintenance jobs
• MTBF
• Unit setup and adjustment time
• Number of minor stoppages
• Number of machines per person</td><td>C = Cost
Manpower reduction rate (personnel reduction rate)
• Maintenance cost reduction rate
• Spare parts cost reduction rate
• Energy cost reduction rate
• Downtime losses</td></tr>
<tr><td>D = Delivery
Delivery delay rate
• Inventory warehousing time
• Stock turnover rate</td></tr>
<tr><td>S = Safety
Number of accidents requiring shutdown</td></tr>
<tr><td>Q = Quality
Defect/rework rate
• Number of delivery claims
• C_p trend</td><td>M = Morale
Number of improvement suggestions
• Number of small group meetings</td></tr>
</table>

Table 9-1. Effectiveness Indices (PQCDSM)

SUMMARY

Each rate or index used to measure TPM effectiveness has advantages and disadvantages; for example, some indices cannot express all the results of activities in certain areas. Maintenance department efficiency indices, in particular, will differ according to the scale of the company and the configuration of its equipment. The effectiveness indices developed thus far should be used with the understanding that they are subject to certain limitations.

CASE STUDIES

Tables 9-2 and 9-3 illustrate case studies at company T and company F respectively.

	PM Management Measure	Details of Measure	Results	Eval.	Remarks
Costs	Cost reduction	Amount of cost reduction	3.2 times	◎	Significant results achieved through productivity improvement and energy-saving activities, etc.
	Plant productivity (companywide average)	$\frac{\text{Standard time}}{\text{Actual time}} \times 100$	1.27 times	◎	Plant productivity was increased through cumulative reductions in downtime, equipment rationalization, work improvements, etc.
	〃 (Komaki plant)	〃	1.32 times	◎	
	Energy reduction per benchmark (kcal/kg of rubber) (whole company)	Trend compared with energy reduction per benchmark (total cost for second half of 1977 set as benchmark)	Down 33%	◎	Significant results achieved through activities of energy-saving taskforce
	Energy reduction per benchmark (Komaki plant)	〃	Down 38%	◎	
	Overall maintenance cost rate	$\frac{\text{Repair + labor + downtime (costs)}}{\text{Total fixed assets}} \times 100$	Down 15%	○	Maintenance costs increased in early stages as a result of equipment restoration work but have fallen continuously since the second half of 1979.
	Stock turnover rate	$\frac{\text{Sales for this year}}{\frac{1}{2}\text{(Leftover stock − this year + last year)}}$	1.4 times	○	Stock levels are continuing to decrease as a result of reductions in breakdowns and setup times, equipment improvements, etc.
	Amount of fixed assets disposed of	Amount of fixed assets disposed of	⅓	◎	Disposal of plant and equipment reduced through extending equipment lifetimes and increasing accuracy of operating plans and design.

	PM Management Measure	Details of Measure	Results	Eval.	Remarks
Equipment Effectiveness	Equipment breakdowns	Number of sporadic breakdowns repaired by maintenance personnel	1/25	◎	Breakdowns and downtime due to breakdowns for machines and dies greatly decreased through PM activities involving everyone from top management to frontline operators; attitudes of all personnel transformed
	Die failures	Number of sporadic die breakdowns	1/6	◎	
	Equipment failure frequency rate	$\frac{\text{Equipment breakdowns}}{\text{Loading time}} \times 100$	1/16	◎	
	Equipment failure rate	$\frac{\text{Stoppage time due to breakdowns}}{\text{Loading time}} \times 100$	1/7	◎	
	Downtime	Total downtime due to breakdowns, setup, cleaning, quality problems, idling, etc.	Down 57%	◎	Breakdowns and idling were reduced and setup procedures improved
	Operating rate	$\frac{\text{Operating time}}{\text{Loading time}} \times 100$	Up 6.2%	◎	Operating rates were improved and targets achieved
	Defects in process	$\frac{\text{Defects in process cost (scrap)}}{\text{Total process output}} \times 100$	Down 45%	○	Equipment-generated quality defects are decreasing as a result of breakdown reductions and increased equipment precision
	Claims rate	$\frac{\text{Claims for period}}{\text{Claims for latter half of 1978}} \times 100$	Down 36%	○	Autonomous maintenance Step 6 and implementation of MQ (machine-quality) control are having particularly beneficial effects

Evaluation code:

◎ = target achieved, outstanding results since introduction of TPM

○ = just short of target, results better than before the introduction of TPM

= target not achieved, little change since introduction of TPM

X = target not achieved, situation better before introduction of TPM

Table 9-2A. Case Study at Company T

	PM Management Measure	Details of Measure	Results	Eval.	Remarks
Product Quality	Accident frequency rate	$\frac{\text{Total accidents}}{\text{Total working hours}} \times 10^6$	Down 70%	○	Number of accidents greatly reduced through better equipment maintenance and housekeeping and increased awareness of all personnel
	Absentee rate	$\frac{\text{Working days lost (absences)}}{\text{Total cumulative working hours}} \times 1000$	Has improved	○	The absence rate target was not achieved, but the rate was lower than the 0.27 national average for the rubber industry in 1978
Safety	State of TPM group activities	$\frac{\text{Total hours in group activities}}{\text{Number of groups}} \times 100$	Promoting activities is easier with new followup system	◎	Each group has spent an average of 6 hours per month on activities over a 30-month period; group activity can be described as vigorous
		$\frac{\text{Frequency of group activities}^{\star}}{\text{Number of groups}} \times 100$ ⋆Running total		◎	Group activity is lively, with a monthly average of 2.7 activities or meetings per group
Education/Morale	Completion of foremen's improvement themes	Running total of themes completed since June, 1978	The foremen's improvement program, started before the introduction of TPM, has continued strongly	○	Four targets per person are set annually; targets have been achieved every year
	State of foremen's presentations	Number of presentations since since June 1978		○	The numbers of foremen's meetings and presentations are continuing close to plan; content is satisfactory, and the program is raising foremen's abilities and improving the workplace
	Improvement suggestions awards (whole company)	Number of suggestions accepted and and rewarded by suggestion committee	6.3 per person	◎	The suggestion program has been popular since TPM was first introduced; the number of suggestions has increased dramatically
	Komaki plant	"	5.4 per person	◎	

Table 9-2B. Case Study at Company T

Item	Index \ Position	Plant manager	Section manager	Foreman	Group leader	Equipment improvement	Autonomous maintenance	Remarks
Production	Productivity	○	○	○				Increase rate
Quality	(Quality product rate)							
	Loss reduction	○	○	○	○	○	○	By individual machine
	Rework reduction		○	○	○	○	○	″
	Claims	○	○	○				Number
Delivery	Daily production target-achievement rate	○	○	○	○			Sample delivery time achievement rate
	Delivery time target-achievement ratio	○	○					
	Inventory reduction	○	○	○				
Safety & Environment	Worker accidents	○	○	○	○			Measurement at fixed intervals and locations
	Workplace noise index		○					
Equipment	Overall equipment effectiveness	○	○	○				
	Availability (operating rate)	○	○	○		○		
	(Equipment breakdowns)	○	○	○	○	○	○	Breakdown rate
	(Equipment breakdown time)		○	○	○	○	○	
	(Setup time)		○	○	○	○	○	Reduction in setup time
	(Worker idling time)							
	(Material waiting time)		○	○				
	(Tool-changing time)			○	○	○	○	Grindstones, dies, cutting tools
	(Operating speed rate)	○	○	○	○	○		
	(Minor stoppages)			○	○	○	○	

Table 9-3A. Managerial Indices by Position (Company F)

Item	Index \ Position	Plant manager	Section manager	Foreman	Group leader	Equipment improvement	Autonomous maintenance	Remarks
Equipment (cont.)	(Downtime losses)		○	○	○	○	○	
	Quality product rate	○	○	○		○		
	(losses)		○	○	○	○	○	Product precision losses
	(rework)		○	○	○	○	○	
	Overall equipment effectiveness target-achievement rate	○	○	○				(Number of lines with an overall efficiency of at least 85% relative to total number of production lines)
Capacity	TPM group classes	○	○					1st grade, 2nd grade, 3rd grade
	TPM group class attendees		○	○				
	Qualified technicians in company	○	○					
	Textbooks produced	○	○					
	Operating standards produced	○	○	○				
Morale	QC circle themes completed	○	○	○	○			Trend in number of suggestions per person
	Improvement suggestions	○	○	○	○			
Energy (per unit raw material)	Hydraulic fluid consumed			○	○	○	○	
	Electricity costs		○					
	Fuel { City gas		○					
	Fuel { LPG		○					
	Maintenance costs	○	○			○	○	

Table 9-3B. Managerial Indices by Position (Company F)

REFERENCES

Hibi, S. *Measuring PM Effectiveness* (in Japanese). Tokyo: Plant Engineering Association, n.d.

Gōtō, Fumio. *PM Correspondence Course Text* (in Japanese). Tokyo: Japan Management Association, n.d.

Takahashi, Giichi. *Promoting Production Maintenance* (in Japanese). Tokyo: Plant Engineering Association, n.d.

Appendix A
The PM Prize for Outstanding TPM Plants

In Japan, the Distinguished Plant Prize (or PM Prize) is awarded annually to plants that successfully implement TPM. As public attention increases, this award is now as highly coveted as the Deming Prize, which has been awarded for exemplary quality programs and achievements for the past 30 years. The PM prize is awarded by the Japan Institute of Plant Maintenance (JIPM), a key promoter of TPM. A special PM Prize committee within JIPM selects the prizewinners.

The PM Prize has been awarded since 1964. While early prizewinners were selected for their outstanding productive maintenance, implementation of TPM has become a requirement for the PM Prize since 1971, when the Nippondenso Company became the first firm in Japan to win the PM Prize for TPM.

The PM Prize is offered in two categories, for large corporations and for firms with fewer than 1,000 employees and less than ¥ 500 million ($2.2 million) in capital assets. The PM Prize committee examines the concrete results achieved by applicants through TPM implementation. Therefore, progress in areas such as systematization or standardization is disregarded. Selection of winners is based on improvements achieved through proper equipment maintenance, such as increased productivity and

Category / Year	Category 1	Category 2
1971	Nippodenso Mitsubishi Heavy Industries	(none)
1972	Toray Industries (Okazaki plant)	Sumiko I.S.P. Co. (Hama plant)
1973	Toyoda Gōsei	Chūsei Rubber Segawa Chemical Industries Hinoda Rubber Industries Chūo Rubber Industries Suzuki Chemical Industries
1974	(none)	Ichiei Industries Hokusei Rubber Kitano Manufacturing
1975	Fuji Photo Film (Odawara plant)	Toyokuni Industries
1976	Special Prize: Toyoda Gōsei Kawasaki Steel (Chiba plant) Yokohama Rubber (Mishima plant)	Special Prize: Suzuki Chemical Industries Kaiyō Rubber Shiota Kasei Co. Yokoyama Spring
1977	Sumitomo Metal Industries (Wakayama plant) Fuji Photo Film (Fujinomiya plant) Yokohama Rubber (Mie plant) Wakō Chemical Industries	Yamako Co.
1978	Chūō Spring (Hekinan plant)	Anjo Denki
1979	Aisan Industries Toyota Steel Works	(none)
1980	Aichi Steel Works	(none)

Note: Category 2 consists of companies with less than ¥500 million ($2 million) in capital assets and less than 1,000 employees.

Table A. PM Prize-Winning Companies (from 1971)

Category / Year	Category 1	Category 2
1981	Anjo Denki Topy Industries (Kanagawa Plant) Tōkai Rubber Industries (Komaki plant) Tokyū Car Corporation (Osaka plant) Maruyasu Kōgyō (Okazaki plant) Matsushita Electric Industries (Mikuni plant)	Kyōwa Precision Teikei Machine Matsuo Seisakusho Miyama Steel
1982	Aishin Seiki Co. Central Motor Wheel (Toyada plant) Nippon Zeon Co. (Kawasaki, Takaoka, Tokuyama, and Mizushima plants) Fuji Photo Film (Yoshida Minami plant)	Tōhoku Satake Seisakusho

Note: Category 2 consists of companies with less than ¥500 million ($4 million) in capital assets and less than 1,000 employees.

quality, reduced costs, reduced inventory, elimination of accidents, pollution control, and the creation of a favorable work environment. Each year, standards (based on actual results) are going up.

Table A lists the PM Prizewinners since 1971, when TPM implementation became a requirement for consideration. By 1982, 29 companies in category 1 and 22 companies in category 2 had been awarded the PM Prize for their TPM. The dramatic increase in prizewinners after 1981 is noteworthy.

Appendix B Application Procedures for the PM Prize

To be considered for the PM Prize in Japan, a company must compile a "TPM Implementation Report" and submit it to the PM Prize Committee. (An outline of award criteria is provided in Appendix C.) Although compiling this report is a formidable task, it forces companies to reevaluate their stages of TPM implementation. This inquiry may lead to the discovery of hidden weaknesses and result in additional improvements; many prize-winners have described it as the perfect opportunity for a comprehensive survey of their equipment maintenance program.

A company being considered for the PM Prize is evaluated by specialists in the relevant fields who will point out weaknesses in their programs and suggest improvements. Applicants invariably find this helpful because the advice received helps boost future TPM results.

Judging occurs in two stages, the initial screening (based on documents) and a second screening (based on a factory visit). Each year companies submit the "TPM Implementation Report" to the JIPM by the end of May for the first screening in June. When a company has passed this screening, the PM Prize Committee notifies it of the date of the factory visit, which is usually some time in July or August. All results are evaluated by early September. Then, the PM Prize Committee consisting of the JIPM chairperson and other experts selects the winners. The award ceremony is held in late September or early October at the National Conference on Equipment Maintenance.

For Japanese companies, the PM Prize symbolizes a new beginning and a challenge to strive for even greater improvement. Frequently, representatives from PM Prize-winning plants are invited to report on their continuing efforts at lectures and conferences, and JIPM sponsors field trips to their plants. Both of these practices are incentives for continued improvement. Repeated presentation of company results to the public and frequent visitors to the plant help employees strive for higher and more challenging goals.

The award level beyond the PM Prize is the Special (Distinguished) PM Prize. Winners of this prize have eliminated the weaknesses discovered earlier by the PM Prize Committee and developed unique maintenance techniques and equipment technology.

Appendix C
Criteria for Awarding the PM Prize

A. Policies and Goals

1. How do company policies relate to equipment management?
2. Are appropriate methods used in setting equipment management policies and goals? Are priorities set appropriately?
3. Are effective managerial indices and evaluation criteria established?
4. Are long-term and annual plans coordinated?
5. Are company policies and goals understood and accepted by everyone (management and employees)?
6. Is the achievement of policies and goals closely monitored?
7. Are results reflected in the subsequent yearly plans?

B. Organization and Management

1. Are the organization and personnel deployment related to equipment management appropriate?
2. Is the TPM-promotion organization effective?
3. Is the TPM-promotion organization well-integrated within the management structure?

4. Is each department participating in TPM?
5. Are the head office and the plants and branch offices cooperating?
6. Are there any obstacles to the communication and effective use of information?
7. Have good relationships been established with outside contractors for equipment, dies, jigs, tools, maintenance work, and so on?

C. Small Group Activities and Autonomous Maintenance

1. Are small groups properly formed?
2. Are small group goals set appropriately?
3. Do groups meet regularly? Are meetings lively?
4. Is the suggestion system active? Are suggestions handled appropriately?
5. How is the achievement of targets confirmed?
6. To what extent are operators performing autonomous maintenance?

D. Education and Training

1. Is TPM understood at every level of the company?
2. Is the scope and level of training appropriate at every level?
3. Is training being carried out according to plan?
4. What is the degree of participation in outside education and training courses?
5. How many employees hold technical licenses and other qualifications?
6. What is the level of knowledge and skill related to maintenance work?
7. Are skills effectively evaluated?
8. Are the effects of education and training being measured?

E. Equipment Management

1. Practice of the Five S's
   - Is equipment free from contamination by dirt, dust, oil, scale, filings, chips, raw materials, and so on?
   - Is action being taken against sources of dirt and other contamination and hard-to lubricate places? Are inspection and cleaning carried out regularly?
   - Are efforts being made to improve visual control by displaying lubrication instructions, maximum and minimum levels on measuring instruments, matchmarks on nuts and bolts, and so on?
   - Are dies, jigs, tools, measuring instruments, cleaning equipment, and raw materials kept clean and in good order?
2. Application of Machine Diagnostic Technology
   Is machine diagnostic technology being used to deal with the following phenomena?
   - cracks, corrosion, looseness, etc.
   - abnormal vibration, noise, temperature, etc.
   - leakage of water, air, steam, gas, oil, etc.
3. Installation Methods
   Are wiring, piping, hydraulic and pneumatic units, electrical control units, and similar equipment installed by suitable methods in appropriate positions?
4. Lubrication
   Are lubricating materials, equipment, and methods and lubricant replenishment and replacement periods selected appropriately? Is lubrication being carried out properly?

F. Maintenance Planning and Management

1. What measures have been taken to assure and increase maintenance quality and efficiency?

2. Have appropriate standards for equipment inspections been set? Are inspections planned and carried out properly?
3. Are annual, monthly, and other maintenance work plans being prepared and executed?
4. Have standards for permanent-stock items, order points, order quantities for spare parts, and other maintenance materials been set? Is stock stored under suitable conditions?
5. Are equipment drawings adequately controlled?
6. Are dies, jigs, tools, and measuring instruments managed properly?
7. Have appropriate record-keeping and data-handling methods been established for equipment deterioration, breakdowns, other stoppages, maintenance manhours, and other aspects of maintenance?
8. Are improvement measures based on data?
9. Are control procedures used properly?

G. Equipment Investment Planning and MP (Maintenance Prevention)

1. Are new-product and process development and new-equipment plans suitably related?
2. Are suitable methods of comparing the economy of equipment investments being used?
3. Are equipment budgets appropriately compiled and controlled?
4. Are MP improvement proposals reflected quickly and accurately in equipment design standards?
5. Are reliability and maintainability considered fully when selecting and designing equipment and its layout?
6. Are equipment testing and acceptance and commissioning control carried out effectively?
7. Is in-house development of equipment dies, jigs, and tools superior?

8. Are measures for preventing the recurrence of serious accidents quick and accurate?
9. Are fixed assets properly managed?

H. Relationship Between Cost Control and Production Quantities, Delivery Times, and Product Quality

1. Is the control of production quantities and delivery times coordinated with equipment management?
2. Are quality control and equipment management coordinated?
3. Are maintenance budgets properly prepared and controlled?
4. Are resource-saving and energy-saving measures adequate?

I. Industrial Safety, Hygiene, and Environment

1. Have satisfactory policies been established related to industrial safety, hygiene, and environmental management?
2. Are management methods and organization appropriate?
3. Is there good coordination between equipment management and industrial safety, hygiene, and environmental management?
4. What are the overall results of the management program?
5. Does environmental management comply with statutory regulations?

J. Results and Evaluation

1. How are results being measured?
2. Is the achievement of policies and goals satisfactory?
3. Is maintenance valued highly from the standpoint of increasing productivity and other general management considerations?
4. Are results reported regularly to academic and professional organizations and in other ways?
5. Have current problems been identified?
6. Is the next TPM promotion plan in place?

Author Profiles

SEIICHI NAKAJIMA

Vice Chairman, Japan Institute of Plant Maintenance
Executive Vice President, Japan Management Association

Seiichi Nakajima graduated with a degree in mechanical engineering from Kanazawa Technical College in 1939. He joined the Japan Management Association in 1949 and since then has served as a management consultant to more than 100 companies. Mr. Nakajima introduced PM to Japan in 1951 and has remained its leading advocate and educator over the past 30 years.

His many publications include *Promoting Equipment Maintenance, An Introduction to Plant Engineering* (in Japanese), and *Introduction to TPM*.

KUNIO SHIROSE

Senior Consultant, Japan Management Association

Kunio Shirose graduated with a degree in applied chemistry from Hokkaido University in 1957 and joined the staff of the Japan Management Association in 1960. His consulting work has focused on improving quality in man-machine systems and increasing equipment efficiency. He serves as TPM adviser to more than 30 companies in Japan, including a number of PM prizewinners, and has co-authored a textbook on process control.

FUMIO GŌTŌ

Chief Consultant, Japan Management Association

Fumio Gōtō graduated with a degree in control engineering from Keiō University in 1969 and joined the Japan Management Association in 1970. His consulting specialties are industrial engineering and quality control. Gōtō has been a TPM advisor since 1973 and has established a particularly solid reputation on his guidance in autonomous maintenance. He now advises more than 30 companies, including prizewinners Tōkai Rubber Industries and Daihatsu Industries.

AINOSUKE MIYOSHI

Chief Consultant, Japan Management Association

Ainosuke Miyoshi graduated with a degree in industrial management from Kanagawa University in 1963 and joined the staff of the Japan Management Association in the same year. Since 1973, his consulting practice has included TPM as well as industrial engineering and materials handling, and in that capacity he serves more than 30 companies. His publications include a book on new-materials handling.

MASAMITSU ASŌ

Maintenance Supervisor, Nippon Steel
Senior Instructor, Japan Institute of Plant Maintenance

Masamitsu Asō joined Nippon Steel Corporation in 1956 and has served continuously as mechanical maintenance work supervisor for over 30 years. In recent years, Mr. Asō has also served as a senior instructor for the Japan Institute for Plant Maintenance. In that capacity, Mr. Asō has worked with many leading companies in Japan, including Toyota Auto Body, INAX, Konica Corporation, Kayaba Industries, and Kubota, Ltd.

Index

Books Available from Productivity Press

Productivity Press publishes and distributes materials on productivity, quality improvement, and employee involvement for business and industry, academia, and the general market. Many products are direct source materials from Japan that have been translated into English for the first time and are available exclusively from Productivity. Supplemental services include conferences, seminars, in-house training programs, and industrial study missions. Send for free book catalog.

Introduction to TPM
Total Productive Maintenance

by Seiichi Nakajima

Total Productive Maintenance (TPM) combines the American practice of preventive maintenance with the Japanese concepts of total quality control (TQC) and total employee involvement (TEI). The result is an innovative system for equipment maintenance that optimizes effectiveness, eliminates breakdowns, and promotes autonomous operator maintenance through day-to-day activities. This book summarizes the steps involved in TPM and provides case examples from several top Japanese plants.
ISBN 0-915299-23-2 / 168 pages / $39.95 / Order code ITPM-BK

Total Productive Maintenance: Maximizing Productivity and Quality

Japan Management Association

Introduce TPM to your work force in this accessible two-part audio visual program, which explains the rationale and basic principles of TPM to supervisors, group leaders, and workers. It explains five major developmental activities of TPM, includes a section on equipment improvement that focuses on eliminating chronic losses, and describes an analytical approach called PM Analysis to help solve problems that have complex and continuously changing causes. (Approximately 45 minutes long.)
167 Slides / ISBN 0-915299-46-1 / $749.00 / Order code STPM-BK
2 Videos / ISBN 0-915299-49-6 / $749.00 / Order code VTPM-BK

Productivity Press, Inc., Dept. BK, P.O. Box 3007, Cambridge, MA 02140 1-800-274-9911

JIT Factory Revolution

Hiroyuki Hirano/JIT Management Library

Here at last is the first-ever encyclopedic picture book of JIT. Using 240 pages of photos, cartoons, and diagrams, this unprecedented behind-the-scenes look at actual production and assembly plants shows you exactly how JIT looks and functions. It shows you how to set up each area of a JIT plant and provides hundreds of useful ideas you can implement. If you've made the crucial decision to run production using JIT and want to show your employees what it's all about, this book is a must. The photographs, from various Japanese production and assembly plants, provide vivid depictions of what work is like in a JIT environment. And the text, simple and easy to read, makes all the essentials crystal clear.
ISBN 0-915299-44-5 / 240 pages / illustrated / $49.95 / Order code JITFAC-BK

The Improvement Book
Creating the Problem-Free Workplace

by Tomo Sugiyama

A practical guide to setting up a participatory problem-solving system in the workplace. This book provides clear direction for starting a problem-free engineering program, a full introduction to basic concepts of industrial housekeeping (known in Japan as 5S), two chapters of examples that can be used in small group training activities, and a workbook for individual use. Informal, using many anecdotes and examples, this book provides a proven fundamental approach to problem solving for any industrial setting.
ISBN 0-915299-47-X / 240 pages / $49.95 / Order code IB-BK

Kanban and Just-In-Time at Toyota
Management Begins at the Workplace (rev.)

edited by the Japan Management Association, translated by David J. Lu

Based on seminars developed by Taiichi Ohno and others at Toyota for their major suppliers, this book is the best practical introduction to Just-In-Time available. Now in a newly expanded edition, it explains every aspect of a "pull" system in clear and simple terms — the underlying rationale, how to set up the system and get everyone involved, and how to refine it once it's in place. A groundbreaking and essential tool for companies beginning JIT implementation.
ISBN 0-915299-48-8 / 224 pages / $29.95 / Order code KAN-BK

Non-Stock Production
The Shingo System for Continuous Improvement

by Shigeo Shingo

Shingo, whose work at Toyota provided the foundation for JIT, teaches how to implement non-stock production in your JIT manufacturing operations. The culmination of his extensive writings on efficient production management and continuous improvement, his latest book is an essential companion volume to his other books on other key elements of JIT, including SMED and Poka-yoke.
ISBN 0-915299-30-5 / 480 pages / $75.00 / Order code NON-BK

BOOKS AVAILABLE FROM PRODUCTIVITY PRESS

Buehler, Vernon M. and Y.K. Shetty (eds.). **Competing Through Productivity and Quality**
ISBN 0-915299-43-7 / 1989 / 576 pages / $39.95 / order code COMP

Christopher, William F. **Productivity Measurement Handbook**
ISBN 0-915299-05-4 / 1985 / 680 pages / $137.95 / order code PMH

Ford, Henry. **Today and Tomorrow**
ISBN 0-915299-36-4 / 1988 / 286 pages / $24.95 / order code FORD

Fukuda, Ryuji. **Managerial Engineering: Techniques for Improving Quality and Productivity in the Workplace**
ISBN 0-915299-09-7 / 1984 / 206 pages / $34.95 / order code ME

Hatakeyama, Yoshio. **Manager Revolution! A Guide to Survival in Today's Changing Workplace**
ISBN 0-915299-10-0 / 1985 / 208 pages / $24.95 / order code MREV

Hirano, Hiroyuki. **JIT Factory Revolution: A Pictorial Guide to Factory Design of the Future**
ISBN 0-915299-44-5 / 1989 / 218 pages / $49.95 / order code JITFAC

Japan Human Relations Association (ed.). **The Idea Book: Improvement Through TEI (Total Employee Involvement)**
ISBN 0-915299-22-4 / 1988 / 232 pages / $49.95 / order code IDEA

Japan Management Association (ed.). **Kanban and Just-In-Time at Toyota: Management Begins at the Workplace** (Revised Ed.), Translated by David J. Lu
ISBN 0-915299-48-8 / 1989 / 224 pages / $34.95 / order code KAN

Japan Management Association and Constance E. Dyer. **The Canon Production System: Creative Involvement of the Total Workforce**
ISBN 0-915299-06-2 / 1987 / 251 pages / $36.95 / order code CAN

Karatsu, Hajime. **Tough Words For American Industry**
ISBN 0-915299-25-9 / 1988 / 178 pages / $24.95 / order code TOUGH

Karatsu, Hajime. **TQC Wisdom of Japan: Managing for Total Quality Control**, Translated by David J. Lu
ISBN 0-915299-18-6 / 1988 / 136 pages / $34.95 / order code WISD

Lu, David J. **Inside Corporate Japan: The Art of Fumble-Free Management**
ISBN 0-915299-16-X / 1987 / 278 pages / $24.95 / order code ICJ

Mizuno, Shigeru (ed.). **Management for Quality Improvement: The 7 New QC Tools**
ISBN 0-915299-29-1 / 1988 / 318 pages / $59.95 / order code 7QC

Monden, Yashuhiro and Sakurai, Michiharu. **Japanese Management Accounting: A World Class Approach to Profit Management**
ISBN 0-915299-50-X / 1989 / 512 pages / $49.95 / order code JMACT

Nakajima, Seiichi. Introduction to TPM: Total Productive Maintenance
ISBN 0-915299-23-2 / 1988 / 149 pages / $39.95 / order code ITPM

Nikkan Kogyo Shimbun, Ltd./Factory Magazine (ed.). **Poka-yoke: Improving Product Quality by Preventing Defects**
ISBN 0-915299-31-3 / 1989 / 288 pages / $59.95 / order code IPOKA

Ohno, Taiichi. **Toyota Production System: Beyond Large-Scale Production**
ISBN 0-915299-14-3 / 1988 / 163 pages / $39.95 / order code OTPS

Ohno, Taiichi. **Workplace Management**
ISBN 0-915299-19-4 / 1988 / 165 pages / $34.95 / order code WPM

Ohno, Taiichi and Setsuo Mito. **Just-In-Time for Today and Tomorrow**
ISBN 0-915299-20-8 / 1988 / 208 pages / $34.95 / order code OMJIT

Psarouthakis, John. **Better Makes Us Best**
ISBN 0-915299-56-9 / 1989 / 112 pages / $16.95 / order code BMUB

Shingo, Shigeo. **Non-Stock Production: The Shingo System for Continuous Improvement**
ISBN 0-915299-30-5 / 1988 / 480 pages / $75.00 / order code NON

Shingo, Shigeo. **A Revolution In Manufacturing: The SMED System**, Translated by Andrew P. Dillon
ISBN 0-915299-03-8 / 1985 / 383 pages / $65.00 / order code SMED

Shingo, Shigeo. **The Sayings of Shigeo Shingo: Key Strategies for Plant Improvement**, Translated by Andrew P. Dillon
ISBN 0-915299-15-1 / 1987 / 208 pages / $36.95 / order code SAY

Shingo, Shigeo. **A Study of the Toyota Production System from an Industrial Engineering Viewpoint** (Revised Ed.)
ISBN 0-915299-17-8 / 1989 / 352 pages / $39.95 / order code STREV

Shingo, Shigeo. **Zero Quality Control: Source Inspection and the Poka-yoke System**, Translated by Andrew P. Dillon
ISBN 0-915299-07-0 / 1986 / 328 pages / $65.00 / order code ZQC

Shinohara, Isao (ed.). **New Production System: JIT Crossing Industry Boundaries**
ISBN 0-915299-21-6 / 1988 / 224 pages / $34.95 / order code NPS

Sugiyama, Tomō. **The Improvement Book: Creating the Problem-free Workplace**
ISBN 0-915299-47-X / 1989 / 320 pages / $49.95 / order code IB

Tateisi, Kazuma. **The Eternal Venture Spirit: An Executive's Practical Philosophy**
ISBN 0-915299-55-0 / 1989 / 208 pages / $19.95 / order code EVS

Audio-Visual Programs

Japan Management Association. **Total Productive Maintenance: Maximizing Productivity and Quality**
ISBN 0-915299-46-1 / 167 slides / 1989 / $749.00 / order code STPM
ISBN 0-915299-49-6 / 2 videos / 1989 / $749.00 / order code VTPM

Shingo, Shigeo. **The SMED System**, Translated by Andrew P. Dillon
ISBN 0-915299-11-9 / 181 slides / 1986 / $749.00 / order code S5
ISBN 0-915299-27-5 / 2 videos / 1987 / $749.00 / order code V5

Shingo, Shigeo. **The Poka-yoke System**, Translated by Andrew P. Dillon
ISBN 0-915299-13-5 / 235 slides / 1987 / $749.00 / order code S6
ISBN 0-915299-28-3 / 2 videos / 1987 / $749.00 / order code V6

TO ORDER: Write, phone, or fax Productivity Press, Dept. BK, P.O. Box 3007, Cambridge, MA 02140, phone 1-800-274-9911, fax 617-868-3524. Send check or charge to your credit card (American Express, Visa, MasterCard accepted).

U.S. ORDERS: Add $4 shipping for first book, $2 each additional. CT residents add 7.5% and MA residents 5% sales tax.

FOREIGN ORDERS: Payment must be made in U.S. dollars (checks must be drawn on U.S. banks). For Canadian orders, add $10 shipping for first book, $2 each additional. For orders to other countries write, phone, or fax for quote and indicate shipping method desired.

NOTE: Prices subject to change without notice.

— Apr. —

UTAH STATE UNIVERSITY PARTNERS PROGRAM

Shigeo Shingo Prize for Manufacturing Excellence

announces the

Shigeo Shingo Prizes for Manufacturing Excellence

Awarded for Manufacturing Excellence Based on the Work of Shigeo Shingo

for North American Businesses, Students and Faculty

ELIGIBILITY

Businesses: Applications are due in late January. They should detail the quality and productivity improvements achieved through Shingo's manufacturing methods and similar techniques. Letters of intent are required by mid-November of the previous year.

Students: Applicants from accredited schools must apply by letter before November 15, indicating what research is planned. Papers must be received by early March.

Faculty: Applicants from accredited schools must apply by letter before November 15, indicating the scope of papers planned, and submit papers by the following March.

CRITERIA

Businesses: Quality and productivity improvements achieved by using Shingo's Scientific Thinking Mechanism (STM) and his methods, such as Single-Minute-Exchange of Die (SMED), Poka-yoke (defect prevention), Just-In-Time (JIT), and Non-Stock Production (NSP), or similar techniques.

Students: Creative research on quality and productivity improvements through the use and extension of Shingo's STM and his manufacturing methods: SMED, NSP, and Poka-yoke.

Faculty: Papers publishable in professional journals based on empirical, conceptual or theoretical applications and extensions of Shingo's manufacturing methods for quality and productivity improvements: SMED, Poka-yoke, JIT, and NSP.

PRIZES

Awards will be presented by Shigeo Shingo at Utah State University's annual Partners Productivity Seminar, held in April in Logan, Utah.

Five graduate and five undergraduate student awards of $2,000, $1,500, and $1,000 to first, second, and third place winners, respectively, and $500 to fourth and fifth place winners.

Three faculty awards of $3,000, $2,000 and $1,000, respectively.

Six Shigeo Shingo Medallions to the top three large and small business winners.

SHINGO PRIZE COMMITTEE

Committee members representing prestigious business, professional, academic and governmental organizations worldwide will evaluate the applications and select winners, assisted by a technical examining board.

Application forms and contest information may be obtained from the Shingo Prize Committee, College of Business, UMC 3521, Utah State University, Logan, UT, 84322, 801-750-2281. All English language books by Dr. Shingo can be purchased from the publisher, Productivity Press, P.O. Box 3007, Cambridge, MA 02140: call 1-800-274-9911 or 617-497-5146.

Japan's "Dean of Quality Consultants"

Dr. Shigeo Shingo is, quite simply, the world's leading expert on improving the manufacturing process. Known as "Dr. Improvement" in Japan, he is the originator of the Single-Minute Exchange of Die (SMED) concept and the Poka-yoke defect prevention system and one of the developers of the Just-In-Time production system that helped make Toyota the most productive automobile manufacturer in the world. His work now helps hundreds of other companies worldwide save billions of dollars in manufacturing costs annually.

The most sought-after consultant in Japan, Dr. Shingo has trained more than 10,000 people in 100 companies. He established and is President of Japan's highly-regarded Institute of Management Improvement and is the author of numerous books, including *Revolution in Manufacturing: The SMED System* and *Zero Quality Control: Source Inspection and the Poka-yoke System*. His newest book, *Non-Stock Production*, concentrates on expanding U.S. manufacturers' understanding of stockless production.

Dr. Shingo's genius is his understanding of exactly why products are manufactured the way they are, and then transforming that understanding into a workable system for low-cost, high-quality production. In the history of international manufacturing, Shingo stands alongside such pioneers as Robert Fulton, Henry Ford, Frederick Taylor, and Douglas McGregor as one of the key figures in the quest for improvement.

His world-famous SMED system is known as "The Heart of Just-In-Time Manufacturing" for (1) reducing set-up time from hours to minutes; (2) cutting lead time from months to days; (3) slashing work-in-progress inventory by up to 90%; (4) involving employees in team problem solving; (5) 99% improvement in quality; and (6) 70% reduction in floor space.

Shigeo Shingo has been called the father of the second great revolution in manufacturing.

— Quality Control Digest

The money-saving, profit-making ideas... set forth by Shingo could do much to help U.S. manufacturers reduce set-up time, improve quality and boost productivity ... all for very little cash.

Tooling & Production Magazine

When Americans think about quality today, they often think of Japan. But when the Japanese think of quality, they are likely to think of Shigeo Shingo,... architect of Toyota's now famous production system.

Boardroom Report

Shingo's visit to our plant was significant in making breakthroughs in productivity we previously thought impossible. The benefits... are more far-reaching than I ever anticipated.

Gifford M. Brown, Plant Mgr.
Ford Motor Company